Palgrave Studies in Literature, Science and Medicine

Palgrave Studies in Literature, Science and Medicine is an exciting, prize-winning series that focuses on one of the most vibrant and interdisciplinary areas in literary studies: the intersection of literature, science and medicine. Comprised of academic monographs, essay collections, and Palgrave Pivot books, the series will emphasize a historical approach to its subjects, in conjunction with a range of other theoretical approaches. The series will cover all aspects of this rich and varied field and is open to new and emerging topics as well as established ones.

Alanna Skuse

Hurt Feelings

Wounding Oneself in Early Modern Literature

palgrave
macmillan

Alanna Skuse
Department of English Literature
University of Reading
Reading, UK

ISSN 2634-6435 ISSN 2634-6443 (electronic)
Palgrave Studies in Literature, Science and Medicine
ISBN 978-3-032-04564-5 ISBN 978-3-032-04565-2 (eBook)
https://doi.org/10.1007/978-3-032-04565-2

Cover illustration: Timacoch / Getty Images

This Palgrave Macmillan imprint is published by the registered company Springer Nature Switzerland AG.
The registered company address is: Gewerbestrasse 11, 6330 Cham, Switzerland

Note on Spelling

In most cases, the original text of primary sources has been maintained, but with some modernisations. 'U' is modernised to 'V' and vice versa, and thorns (e.g. y^t, y^m) expanded (e.g. to 'that', 'them'). Capitalisations have also been modernised, except in those cases where the capitalisation of a word seems designed to confer emphasis.

Acknowledgements

This work was generously funded by a Wellcome Trust University Award, grant number 218099/Z/19/Z.

My research has been supported by academic colleagues, including Professor Michelle O'Callaghan, Professor Andrew Mangham, Dr Amie Bolissian, Dr Ella Sbaraini, Professor Tina O'Connell and members of the Centre for Health Humanities, the Early Modern Research Centre and the 'Identities' research theme group at the University of Reading. Work from the chapters on self-gelding and autoglossotomy appeared in *Journal of the Social History of Medicine* and *Journal of the Society of Renaissance Studies* respectively, and I thank the readers of those articles for their suggestions.

The greatest thanks for this project goes to the non-academic partners who illuminated new ways of thinking about this topic: staff at Self Injury Support, especially Jen Price and Murphy Pickard; the staff of Acta Theatre Company; freelance theatre facilitator Hannah Wood; and the participants of the art and theatre workshops, who showed up with compassion and creativity.

Competing Interests The author has no competing interests to declare that are relevant to the content of this manuscript.

Author's Note

This book deals with difficult topics of self-injury, suicide, sexual violence and mental ill health. In the United Kingdom, support for people dealing with these issues can be found through your GP or via the following organisations:

Self-Injury Support www.selfinjurysupport.org.uk / 0808 800 8088
Mind www.mind.co.uk / 0300 123 3393
CALM www.thecalmzone.net / 0800 585858
Papyrus www.papyrus-uk.org / 0800 068 41 41
Samaritans www.samaritans.org / 116 123

Contents

CHAPTER 1

Introduction: Why Self-Wounding?

1592. A man who has committed murderous revenge on the people who killed his son is apprehended for his crimes. Facing torture, he bites out his tongue before killing himself.
1599. A young wife is convinced by her husband's strange behaviour that he is keeping a secret. To convince him that she can be trusted with the information, she shows him a cut she has made in her own leg.
1684. A man convinced that his wife is being unfaithful decides to test his theory by cutting off his own testicles. Should she bear further children, he says, then he'll know he is not the father.
1685. A soldier returned from the wars has a quarrel with his mother. In a fit of rage, he stabs himself in the stomach.

In early modern England, stories of people hurting themselves abounded. Some were fictional, some apparently factual, though refracted through the storyteller's lens. All raised questions for those who witnessed the acts in person or through the page: What can a wound do? What circumstances does it make visible or invisible? How does it change the way that people relate to one another? What power over the body does it confer or abdicate? This book seeks to answer those questions for the early modern period and examine the little-studied phenomenon of self-wounding in detail for the first time. In so doing, it also raises broader questions about how we approach the actions of historical actors when they feel close to home. Can we bring early modern narratives to bear on modern experiences and vice versa? And why should we want to do so?

A. Skuse, *Hurt Feelings*, Palgrave Studies in Literature, Science and Medicine, https://doi.org/10.1007/978-3-032-04565-2_1

While interpersonal violence in the early modern period has been the subject of much scholarship, the hurts people inflicted on themselves have been little thought of. Yet self-wounding took place in all sorts of ways and was narrated in all sorts of places. Medical texts gave accounts of men who castrated themselves to shock nagging wives. Plays showed self-inflicted cuts and stabs, calculated to move audiences both on- and off-stage. Poetry told how grieving women tore out their hair and scratched their faces, and tales of religious self-mutilation struck an uneasy balance between spiritual edification and sinful self-destruction. The following chapters argue that when we examine self-injury we witness a practice which was politically, socially and symbolically active as well as personally affective, with both aesthetic and somatic dimensions. Whereas accounts of pain suffered accidentally, through illness, or at the hands of others show how people chose to interpret what happened *to* them, stories about self-inflicted wounds show how people chose to embrace pain for a range of ideological or pragmatic reasons. It was, I will argue, readable through a variety of frameworks, of which the medical was among the least important. People who injured themselves might be mentally ill, but they were more likely to be protesting an injustice, proving their authenticity, managing their emotions or asserting their independence. Though these acts were uncommon, they were imaginatively and socially powerful. Their study tells us much about how early modern people related to their bodies and emotions, and suggests new directions for investigations in early modern rhetoric and embodiment.

1.1 Parameters

What is 'self-wounding'? That the acts described in this book comprise a group of practices that can and should be studied together is not a straightforward proposition. The early modern lexicon has no terminology for self-wounding, meaning that individual instances of, say, self-stabbing, self-gelding and auto-amputation appear as disparate acts. Part of this project's purpose is thus to demonstrate how we might study such fluidly related phenomena by tracing both unity and variety among self-wounding events. Different kinds of self-inflicted injury were united by common languages and scenarios. For example, I show in Chap. 3 that the fictional autoglossotomy of *The Spanish Tragedy*'s Hieronimo drew much of its meaning from classical stories in which tongue-biting was employed as a means of defiance in the face of tyranny. Though the self-gelding incidents described in Chap. 5 are a different kind of action, they share with Hieronimo's autoglossotomy both the potential to be read as forms of resistance, and a narrative structure (setting – discovery – exposition) which allows them to appear as dramatic interludes within medical and scientific texts. Drew Daniel's model of 'assemblage' is particularly apposite here. Writing on early modern melancholy, he describes that condition as determined rather like we determine the difference between 'some grains' and 'a heap of grain'—not a single symptom but the accrual of symptoms from a recognised range.[1] Open-ended and hybrid, this 'epistemological-affective assemblage' is ragged at the edges and fluid in its composition.[2] Similarly, self-wounding and its effects form a constellation rather than a neatly defined field. Accounts vary tonally as they appear in a wide range of texts: imaginative and instructive, serious and jovial. This book includes stories about and representations of self-wounding from plays, poetry, ballads, medical textbooks, histories, martyrologies and miscellanies. Self-wounding acts also differ in modality and severity, from face-scratching and hair-tearing to amputating one's genitals or biting out one's own tongue.

What is excluded is also important. Though we might now be horrified to see somebody opening a vein or applying caustic chemicals to the skin, for early modern people these were medicinal practices which shared little with the other kinds of self-wounding I examine. Extensive scholarship has shown that the painfulness of such acts was sometimes taken as proof of their effectiveness (and conversely, painful experiences might be construed as medicinal).[3] Many early modern patients and medical practitioners seemingly embraced the maxim of 'if it's hurting, it's working', as much by necessity as choice.[4]

[1] Daniel Drew, *The Melancholy Assemblage: Affect and Epistemology in the English Renaissance* (Fordham University Press, 2013), 27.

[2] Daniel, *The Melancholy Assemblage*, 28.

[3] See for example Jenny Mayhew, "Godly Beds of Pain: Pain in English Protestant Manuals (ca. 1550–1650)" and Jan Frans van Dijkhuizen, "Partakers of Pain: Religious Meanings of Pain in Early Modern England", both in *The Sense of Suffering: Constructions of Physical Pain in Early Modern Culture,* edited by Jan Frans van Dijkhuizen and Karl A. E. Enenkel (Brill, 2008).

[4] Nonetheless, deliberate self-injury can add still further to our understanding of medicinally inflicted pain by serving as an extreme example of how taking control of one's bodily experience could be understood as an act of personal agency.

This book also excludes a broader field of intentionally harmful activities which might linger at the edges of our constellation, such as self-starvation, object-swallowing or self-poisoning. These practices are not treated here partly because, as I discuss below, in-depth work already exists on these topics, and partly because I am interested in the visible wound itself: as the locus of pain, as a tangible proof of the self-wounding act and as a site where the bodily envelope is opened to the wider world. The exception to the rule is Chap. 6, which uses recent work on stories of women's self-starvation to suggest how those accounts might have shaped how a seventeenth-century audience responded to a tale of eighth-century self-mutilation. Most of the acts described in this book penetrate the skin, and in so doing draw attention to the uniqueness of that organ as something which encloses the body and yet is itself a means of sensing which cannot be closed off in the same way as sight, hearing, smell and taste.

My choice of the phrases 'self-wounding' or 'self-injury' to describe the acts in this book is therefore a deliberate one. They are the most accurate (though still imperfect) terms to describe the practices I examine, which for the most part break the skin and leave a physical trace. These slightly awkward expressions also underline the rift between the activities and discourses I trace here and those of modern 'self-harm', a term which came into widespread use in the twentieth century.[5] The relationship between modern acts of self-harm and pre-modern acts which may look the same has been a subject of debate among historians of medicine for some time, with varying results.[6] Armando Favazza's foundational 1987 study of self-harm, *Bodies Under Siege*, approached the phenomenon as a near-universal and transhistorical one. 'Self-mutilation', he argued, 'is a morbid form of self-help', and as such can be found in one form or another across widely

[5] 'Self-harm' is now seldom used by clinicians or sociologists, who prefer the term 'non-suicidal self-injury' (NSSI). However, it remains in widespread use in mainstream media coverage of self-injury, and among people who self-injure.

[6] Issues of terminology are astutely discussed in Drew Daniel's *Joy of the Worm*, which argues that there exists an important difference between 'suicide' (a term coined in the mid-seventeenth century and which describes a morally and spiritually dangerous act) and 'self-killing' (which is open to a broader range of interpretations and presentations): 'simply put, "self-killing" includes a wider array of forms of voluntary death than "suicide," because it is not saddled with overfamiliar connotations of pathology and mood disorder. Early modern literature teems with such an array of forms because it was written before these connotations took hold'. Drew Daniel, *Joy of the Worm: Suicide and Pleasure in Early Modern English Literature*, Thinking Literature (University of Chicago Press, 2022), 7.

varied chronological and geographical settings.[7] Patricia and Peter Adler take a slightly different approach. They argue that from antiquity through the Renaissance period, self-injury of various kinds existed and was the preserve of either 'religious fanatics, leaders, and their followers' or 'outcasts and the severely disturbed or mentally ill'.[8] However, it is only in 'the past several centuries' that self-injury has been defined as a psychological pathology, predominantly belonging to white middle-class young women.[9]

Recent studies have been far more cautious about asserting that self-harm is a universal impulse which is merely named differently in different contexts. Examining the twentieth-century history of self-harm, sociologist Chris Millard insists that 'socially directed attempted suicide *cannot exist* as a concept or pattern of behaviour independent of the institutional channels and professional scrutiny through which it is constituted'.[10] In her *Psyche on the Skin*, Sarah Chaney similarly argues that 'while people had certainly harmed themselves prior to this period, the late nineteenth century was the first time that diverse acts—from skin-picking to amputation—became regarded as equivalent behaviours'.[11] One of very few works on self-injury in the early modern period is Giulio Pertile's insightful article on self-mortification in *King Lear*, in which he considers Edgar/Poor Tom's practice of incising his own arms with 'pins, wooden pricks, nails, sprigs of rosemary' (2.2.187). Pertile uses modern thinking on self-injury to untangle Edgar's claim to be both 'numbed and mortified' by his injuries, relating the opposition of sense-subduing 'numbing' and sense-awakening 'mortification' to studies in which the suppression of difficult emotions and the desire to provoke feeling are both cited as motivations to self-injure.[12] He stops short, however, of making claims about concepts of self-injury as either historically specific or transhistorical.

Despite these difficulties, I argue that there is nonetheless scope for a nuanced story which acknowledges the resemblance between historical

[7] Armando R. Favazza, *Bodies Under Siege: Self-Mutilation and Body Modification in Culture and Psychiatry* (Johns Hopkins University Press, 1987), xix.

[8] Patricia A. Adler and Peter Adler, *The Tender Cut: Inside the Hidden World of Self-Injury* (New York University Press, 2011), 14.

[9] Adler and Adler, *The Tender Cut*, 2.

[10] Chris Millard, *A History of Self-Harm in Britain: A Genealogy of Cutting and Overdosing* (Palgrave Macmillan, 2015), 26. My italics.

[11] Sarah Chaney, *Psyche on the Skin: A History of Self-harm* (Reaktion Books, 2017), 13.

[12] Giulio J. Pertile, "King Lear and the Uses of Mortification". *Shakespeare Quarterly* 67, no. 3 (2016): 319–43. https://doi.org/10.1353/shq.2016.0039, 327.

and modern discourses about self-injury without conflating the two. Indeed, to deny any relationship between historical and modern experiences of self-injury may lead us to unjustly exclude people who self-injure from narratives which involve them, violating the disability studies mantra of 'nothing about us without us'. This is timely; the delicate work of finding a balance between productive and reductive kinds of presentism is of much concern in the health humanities. For instance, in her recent work on voice-hearing in the medieval period and its parallels in modern narratives of psychosis, Hilary Powell notes that the 'similarit[y] is curious and justifies the comparison', but also asks: 'How do we move beyond the superficial small talk of similarities and differences to a more satisfying and deeper engagement which can open up new perspectives and effect some real change?'[13]

In this book, I ponder this question whilst using diverse critical approaches to undertake the historicist work of reconstructing early modern approaches to self-wounding, and thinking about what self-wounding does in literary texts. In the final chapter, I return to reconsider how my analysis of the uses of early modern self-wounding sits alongside modern sociological and psychological understandings of non-suicidal self-injury which have recentred its socio-cultural aspects. I do this partly with reference to projects which have run alongside the research for this book, in which people with lived experience of self-injury have been invited to view those experiences in historical perspective, and to create works of art and theatre based on those reflections. In addition, the words of these participants serve as epigraphs to each chapter (the whole play devised by participants is appended to the end of the book). They function both as an attempt to provoke thought about how historical and modern phenomena might interact, and to re-insert feeling into the sometimes oddly clinical business of literary-historical research.

1.2 Pain and Wounds in Scholarship

In choosing to study *self*-wounding, this book adopts a unique perspective on early modern experiences of the injured and pained body. Work on these topics to date has tended to emphasise a single interpretive

[13] Hillary Powell, Hilary. "Voices in Psychosis: A Medieval Perspective". In *Voices in Psychosis: Interdisciplinary Perspectives*, ed. Ben Alderson-Day, Charles Fernyhough and Angela Woods (Oxford University Press, 2022), 96, 98.

framework, taking (for example) neo-stoic or Puritan philosophy and showing how that lens was applied to a broad range of pain experiences: wounding, illness, grief and so forth.[14] This book reverses that model by taking a relatively discrete variety of pain experience and showing how it acquired different meanings through the application of different cultural frames. In this way, I hope to emphasise the breadth of interpretations which could be applied to pain and wounding at any given time. Additionally, scholarship on both pain and wounding tends to focus on things which happened *to* the wounded or pained person, such as interpersonal violence, accidents and judicial punishments.[15] In such cases, the person in question has implicitly been positioned as an intentional object towards which action has been directed, and much of the work of pain narratives is involved with helping the person-in-pain re-establish themselves as an agential subject (or conversely, preventing them from doing so). By contrast, in self-wounding events, a subject chooses to treat their own body as object, at least in part. To state the obvious, they are at once the doer of violence and the person to whom violence is done. This unique vantage point allows us to examine pain as it is rarely seen, from a proactive rather than reactive perspective. In other words, self-wounding may offer an 'inside-out' look at how early modern people treated pain as a practice that could transform their relationship to the world.

While it breaks new ground, however, this study also builds on a rich historiography around early modern bodily experience in general and pain and wounding in particular. Scholarship of the last three decades has shown that these phenomena were richly theorised in medicine, literature, religion and philosophy. In particular, attention has focussed on interpersonal violence and its representations. Studies by Cynthia Marshall, Katherine Eisamann Maus, Kim Solga and Jonathan Davies among many

[14] See for instance Jan Frans van Dijkhuizen, "Partakers of Pain"; Javier Moscoso, "Learning and Teaching Pain," in *The Routledge History of Emotions in Europe, 1100–1700*, ed. Andrew Lynch and Susan Broomhall (New York: Routledge, 2020), 169–82; Alison Searle, "'A Kind of Agonie in My Thoughts': Writing Puritan and Non-Conformist Women's Pain in 17th-Century England," *Medical Humanities* 44, no. 2 (2018): 125–36, https://doi.org/10.1136/medhum-2017-011407; Mayhew, "Godly Beds of Pain".

[15] Elizabeth Hanson, "Torture and Truth in Renaissance England". *Representations* 34 (1991): 53–84. https://www.jstor.org/stable/2928770; Cynthia Marshall, *The Shattering of the Self: Violence, Subjectivity, and Early Modern Texts* (John Hopkins University Press, 2022); Lisa Silverman, *Tortured Subjects: Pain, Truth, and the Body in Early Modern France* (University of Chicago Press, 2001).

others have shown how both fictive on-stage and real-life state-sanctioned forms of early modern violence wounded the body to excess, layering varieties of harm one atop the other (hanging, drawing and quartering, for example).[16] In so doing, they endlessly reiterated the divisibility of the body. Margaret E. Owens summarises:

> [D]ismemberment and decapitation highlight corporeal semiosis. When a limb is amputated, the detached body part enters a disturbingly liminal state somewhere between subject and object status ... Dismemberment tends to expose the social and political inscription of the human body and hence of the subject.[17]

Fascination with the phenomenology of the body as that which both experiences and is experienced by others is likewise evident in Renaissance preoccupations with the interior of the body as revealed through wounding or dissection. In Matthew Greenfield's analysis of Christopher Marlowe's 'wound knowledge', he suggests that 'Wounded characters ... develop an uncanny knowledge of what is happening inside their bodies, including the precise anatomy of their injuries and the physiology of the onset of death'.[18] In a similar vein, Katherine Rowe describes the early modern preoccupation with anatomies of the hand as an effort to locate agency and subjectivity within the matter of the body, such that 'the hand becomes the prominent vehicle for integrating sacred mystery with corporeal mechanism'.[19] Moreover, dissection and wounding become not only a practical but an imaginative tool for laying bare the 'inside' of a person. Jonathan Sawday's influential *The Body Emblazoned* describes how poetic 'anatomies' sought to expose both object and author, while John V. Nance

[16] Marshall, *The Shattering of the Self*; Katharine Eisaman Maus, "Taking Tropes Seriously: Language and Violence in Shakespeare's *Rape of Lucrece*," *Shakespeare Quarterly* 37, no. 1 (1986): 66–82, https://doi.org/10.2307/2870192; Kim Solga, "Rape's Metatheatrical Return: Rehearsing Sexual Violence among the Early Moderns," *Theatre Journal* 58, no. 1 (2006): 53–72; Jonathan Davies, *Aspects of Violence in Renaissance Europe*, New edition (Ashgate Publishing Ltd, 2013).

[17] Margaret E. Owens, *Stages of Dismemberment: The Fragmented Body in Late Medieval and Early Modern Drama* (University of Delaware Press, 2005), 20.

[18] Matthew Greenfield, "Christopher Marlowe's Wound Knowledge". *PMLA* 119, no. 2 (2004): 233–46. https://doi.org/10.1632/003081204X21289, 233.

[19] Katherine Rowe, "God's Handy Worke". In *The Body in Parts: Fantasies of Corporeality in Early Modern Europe,* ed. David A. Hillman and Carlosa Mazzio (Routledge, 1997), 287.

views Thomas Nashe's preoccupation with violence in *The Unfortunate Traveller* as providing 'more of an anatomy than anatomy itself'.[20]

That emphasis on wounds as sites of discourse is extended and complicated by emerging work on pain. Rejecting Elaine Scarry's contention that physical pain stymies communication, scholars of the early modern period have consistently argued that pain was experienced as having personal, social and religious meaning, and that suffering pain was as likely to help constitute one's identity as it was to threaten it.[21] Kimberley Huth, for instance, eloquently suggests that the sixteenth and seventeenth centuries comprised a liminal period in understandings of pain—after its widespread deployment as a tool of torture, yet before its categorisation as a meaningless physical event in the later eighteenth century. In this space, she argues, emerges 'the possibility that suffering pain (not just inflicting it) can be, in certain ways, intentional ... the real significance of pain is not its ability to disclose the interior but rather its ability to be manipulated according to an individual's intentions regarding his or her outward identity'.[22] Accordingly, scholars have shown that many philosophical and religious texts contained detailed advice about how to approach pain as a means to personal growth, providing scripts with which people in pain could represent their experiences as meaningful to onlookers and to themselves.[23] Alison Searle, for instance, describes how in Puritan texts, women's pain

[20] Jonathan Sawday, *The Body Emblazoned: Dissection and the Human Body in Renaissance Culture* (Routledge, 2013); John V. Nance, "Gross Anatomies: Mapping Matter and Literary Form in Thomas Nashe and Andreas Vesalius". In *The Age of Thomas Nashe*, ed. Stephen Guy-Bray, Joan Pon Linton and Steve Mentz (Routledge, 2013), 124.

[21] Elaine Scarry, *The Body in Pain: The Making and Unmaking of the World* (Oxford University Press, 1988).

[22] Kimberly Huth, "Figures of Pain: Suffering and Selfhood in Early Modern English Literature" (Ph.D., The University of Wisconsin - Madison, 2009), ProQuest, 25.

[23] See Stephen Pender, "Seeing, Feeling, Judging: Pain in the Early Modern Imagination," in *The Sense of Suffering: Constructions of Physical Pain in Early Modern Culture*, ed. Jan Frans van Dijkhuizen and Karl A. E. Enenkel (Brill, 2008), 469–98; Kristine Steenbergh, "Green Wounds: Pain, Anger and Revenge in Early Modern Culture," in *The Sense of Suffering: Constructions of Physical Pain in Early Modern Culture*, ed. Jan Frans van Dijkhuizen and Karl A. E. Enenkel (Brill, 2008), 165–88; Lisa Wynne Smith, "'An Account of an Unaccountable Distemper': The Experience of Pain in Early Eighteenth-Century England and France," *Eighteenth-Century Studies* 41, no. 4 (2008): 459–80; Jan Frans van Dijkhuizen and Karl Enenkel, "Introduction: Constructions of Physical Pain in Early Modern Culture," in *The Sense of Suffering: Constructions of Physical Pain in Early Modern Culture*, ed. Jan Frans van Dijkhuizen and Karl A. E. Enenkel (Brill, 2008), 1–18.

might function as 'both an affective and exegetical form of piety'.[24] Similarly, Jenny Mayhew notes numerous tactics recommended in protestant 'godly handbooks' by which readers could 'manipulate their own understanding of pain' and embrace its 'potential to clarify, discipline, refine and elevate the sinful self'.[25] Importantly, she contends that by using such tactics, 'the faithful appear to have succeeded in changing their actual experience of pain'.[26] Narratives about pain also appeared outside religious discourse. Lisa Wynne Smith has shown how well into the 1700s, humoral models were used by patients and medical practitioners to make sense of pain, despite the emergence of other medical theories such as iatrochemistry and vitalism.[27] Crucially, Smith, in common with Dijkhuizen and Enenkel, argues that descriptions of pain distinguished hazily, if at all, between emotional and physical suffering, with words we might think of as exclusively physical or emotional descriptors being used as something 'more than a metaphor' for pains which were understood to afflict body, mind and soul.[28]

1.3 Emotions, Feelings, Passions

The importance of pain and wounds in early modern systems of thought—their ability to constitute and communicate identity, to build character or annihilate it—depended on the unique intellectual systems within which early modern physical and emotional experience was conceptualised. To an extraordinary degree, these systems emphasised connectedness between individuals, between different aspects of the human experience and between a human body and its environments, all of which I will argue were influential on depictions of self-wounding.

They also fostered a wide-ranging and malleable terminology, which poses a challenge to modern scholars. Contemporary philosophers and sociologists of feeling have sometimes divided socially mediated 'emotion' from pre-social 'affect'. This approach seems out of step with early modern thinking, in which somatic experience was so deeply entwined with

[24] Alison Searle, "'A Kind of Agonie in My Thoughts': Writing Puritan and Non-Conformist Women's Pain in 17th-Century England". *Medical Humanities* 44, no. 2 (2018): 125–36. https://doi.org/10.1136/medhum-2017-011407, 126.

[25] Mayhew, "Godly Beds of Pain", 299, 304.

[26] Mayhew, "Godly Beds of Pain", 299.

[27] Smith, "An Account of an Unaccountable Distemper", 462.

[28] Smith, "An Account of an Unaccountable Distemper", 464.

cognition, and with the exterior world via the Galenic non-naturals, that no experience of feeling could easily be said to be *purely* bodily.[29] More complicated, however, is the matter of early modern words for, and supposed mechanisms of, feeling. Scholars such as Thomas Dixon have argued that 'the emotions did not exist until just under two hundred years ago'.[30] That is, the term 'emotion' was not used in the way we think of it today in the early modern period, and Dixon argues that we ought to use the differentiated categories of 'passions', 'appetites' and 'affections' instead. This is somewhat hampered by the fact that few scholars, modern or early modern, agree on exactly what is meant by each of these terms. For example, 'affections' are said by Elena Carrera and by Louis Charland and R.S. White to be close to the modern term 'emotions' in that they describe passing states of mind rather than longer-term dispositions.[31] However, it is evident from Thomas Wright's 1604 *The Passions of the Minde in Generall* that any delineation between passions and affections was at best partial:

> They are called Passions … because when these affections are stirring in our minds, they alter the humours of our bodies, causing some passion or alteration in them. They are called perturbations, for that … they trouble wonderfully the soule, corrupting the judgement, & seducing the will … They bee also named affections, because the soul by them, either affecteth some good, or for the affection of some good, detesteth some ill. These passions then be certaine internall actes or operations of the soule, bordering upon reason and sense, prosecuting some good thing, or flying some ill thing, causing therewithall some alteration in the body.[32]

[29] Monique Scheer, whose work informs parts of this book, argues that 'affect' is also an unhelpful term in modern settings, because it falsely implies that some feelings are precultural. (Monique Scheer, "Are Emotions a Kind of Practice (and is that what makes them have a history)? A Bourdieuian Approach to Understanding Emotion". *History and Theory 51*, no. 2 (2012): 193–220, 198n.

[30] Thomas Dixon, *From Passions to Emotions: The Creation of a Secular Psychological Category* (Cambridge University Press, 2003), 1.

[31] Louis C. Charland and R.S. White, "Anatomy of a Passion: Shakespeare's The Winter's Tale as Case Study". In *Ordering Emotions in Europe, 1100–1800,* ed. Susan Broomhall (Brill, 2015), 198; Elena Carrera, "Augustinian, Aristotelian, and Humanist Shaping of Medieval and Early Modern Emotion: Affectus, Afectio, and 'affection' as Travelling Concepts." In *Before Emotion: The Language of Feeling, 400–1800,* ed. Michael W. Champion, Kirk Essary, and Juanita Feros Ruys (Routledge, 2019), 187–88.

[32] Thomas Wright, *The Passions of the Minde in Generall. Corrected, Enlarged, and with Sundry New Discourses Augmented* (Valentine Simmes and Adam Islip, 1604), 8.

Wright's work on this subject was probably the clearest and most comprehensive of his era, and even he could or would not sharply distinguish between terms.[33] Moreover, his account of the passions themselves emphasises that while they should primarily be understood as motions of the sensitive part of an Aristotelian tripartite soul (consisting of natural, sensitive and rational faculties), they were entwined with both rational thought and the material humours. '[O]ur Passions', he explained, 'by three wayes are mooved; by humours arising in our bodies, by externall senses and secret passage of sensuall objects, by the descent or commaundement of reason'.[34] Thus, though they proceeded from the mind, and could originate from the rational as well as (more usually) the sensitive faculties, passions made their effects known by provoking the heart to draw to itself corresponding humours (e.g. choleric humours for an angry passion) which then circulated throughout the body, causing physiological effects.

In Chap. 4, I will argue that certain discussions deliberately used the language of 'passions' to differentiate between one's immediate feelings about an event or object and the rounded emotions that resulted from the (conscious and unconscious) use of affective self-management techniques. In general, however, I use the terms 'emotions' or 'feelings'. This is partly because, as Benedict Robinson contends, while 'there are places where the sets of passion and emotion do not overlap', nonetheless 'the concept of emotion emerged within the frame of a concept of passions and as a new way of *construing* passion: the terms exist in a direct historical relation'.[35] Moreover, imprecision is sometimes helpful; dwelling too closely on the finer points of affections, motions and perturbations risks failing to see the wood for the trees. What mattered to most early modern people, I will argue, was a general sense that joys, sorrows, rages and lusts happened at the meeting places of physiology, volition and rhetoric, where 'brain, body and world blow through, and flow leakily and violently into one another'.[36] Though this period has traditionally been billed as the birthplace of the individual, much recent scholarship has shown how the extensive influence

[33] Susan James notes that the interchangeability of terms such as 'passions', perturbations', and so on was remarked upon by Augustine, and 'widely invoked' by early modern authors (Susan James, *Passion and Action: The Emotions in Seventeenth-Century Philosophy* (Oxford University Press, 1999), 11).

[34] Wright, *The Passions of the Minde in Generall*, 149.

[35] Benedict S. Robinson, "Thinking Feeling". In *Affect Theory and Early Modern Texts: Politics, Ecologies, and Form*, ed. Amanda Bailey and Mario DiGangi (Springer, 2017), 17.

[36] Miranda Anderson, *The Renaissance Extended Mind* (Springer, 2015), 103.

of outside factors on the humours, and the acknowledged friction between rational thought and passionate urges, fostered a view of the self as inwardly multitudinous. Writing 'on the inconstancy of our actions', Montaigne concluded from the 'supple variation and easie-yeelding contradiction' observable in himself and his fellow men that 'We goe not, but we are carryed: as things that flote, now gliding gentlie, now hulling violently; according as the water is, either stormy or calme'.[37]

1.4 Making Meaning

I have argued elsewhere that beliefs about the body-mind in this period are characterised by multiplicity and heterogeneity.[38] Stories of self-wounding make this multiplicity evident, sometimes revealing a mind divided against itself, and at others, communicating a firm sense of personal integrity. Accordingly, they demand scholarship which similarly embraces multiple disciplines and modes of thought, reading intersections with later theoretical concepts as well as with contemporary texts and practices. One particularly important way in which this book grapples with the potential of self-wounding to both invite and resist meaning-making is by considering self-injurious acts as 'embodied practice'. I borrow this term from work by the anthropologist Monique Scheer, which in turn builds on Pierre Bourdieu's *Outline of a Theory of Practice.* Bourdieu embraced practice theory as a mode of sociological analysis, a middle ground between top-down objectivist views of social phenomena and phenomenological or subjectivist viewpoints which focussed intently on individual experience.[39] In 'practice', he found a means to describe actions

[37] Michel de Montaigne, "Of the inconstancie of our actions" *Essays Written in French by Michael Lord of Montaigne.* Transl. by John Florio (Melch. Bradwood for Edward Blount and William Barret, 1613), n.p.

[38] Alanna Skuse, *Surgery and Selfhood in Early Modern England: Altered Bodies and Contexts of Identity.* (Cambridge University Press, 2021).

[39] I am acutely aware that I use Bourdieu's concepts here selectively, and far more lightly than he would have intended, in much the same way that early modern studies have tended to embrace 'Foucauldian' analyses of power relations without subscribing to Foucault's epochal view of history. Bourdieu himself suggested that literary criticism (at least of realist literature) could be an exercise in showing sociological structures in action, so that if one ascertained the various forms of capital possessed by all the characters, and their relationship with one another through fields of power, then the narrative of a text was predictable as a series of set-pieces experimenting with different combinations and scenarios. He was, however, also committed to methodological eclecticism and to theory as a tool for use rather than

generated in the interrelationship between the 'habitus' of everyday life and the particular encounters with different kinds of power (economic, cultural and symbolic) which take place at any given time. 'The habitus', he argues, 'produces practices which tend to reproduce the regularities immanent in the objective conditions of the production of their generative principle, while adjusting to the demands inscribed as objective potentialities in the situation'.[40] Bourdieu's theory, which emphasises the soft power exercised by the economically and culturally advantaged over the disadvantaged, may seem an odd fit to the stories of self-wounding as resistance in this book. Nonetheless, as Scheer explains,

> Since the habitus does not dictate the exact course of action in practice but rather provides a 'feel' for the appropriate movements, gestures, facial expressions, pitch of the voice, and so on, it leaves space for behaviors not entirely and always predictable, which can also instantiate change and resistance rather than preprogrammed reproduction. *Social practice is spontaneous while drawing on conventional forms, individual by virtue of a unique montage of embodied social structures.*[41]

Even when it is at its most adversarial to power, self-wounding works because it speaks in the existing language of discourses about the body and communication. Portia's voluntary wound adheres to principles of rhetoric while Hieronimo's tongue-biting repudiates the sovereignty of the state over individual bodies, but they are both enmeshed with their 'habitus', in a way that goes beyond simply influencing or being influenced by it. As Fay Bound Alberti puts it, 'Bourdieu's approach makes mental and bodily experiences of emotion a product of the internalisation of social processes, and social processes as the embodiment of individual dispositions and actions'.[42]

an end in itself. See John R.W. Speller, *Bourdieu and Literature*, 1st ed. (Open Book Publishers, 2011).

[40] Pierre Bourdieu, *Outline of a Theory of Practice* (Cambridge University Press, 1977), 78. See also Timothy Barrett, "Storying Bourdieu: Fragments Towards a Bourdieusian Approach to 'Life Histories'", *International Journal of Qualitative Methods* 14, no.5 (2015); Helle Samuelsen and Vibeke Steffen. "The Relevance of Foucault and Bourdieu for Medical Anthropology". *Anthropology and Medicine* 11, no.1 (2004): 3–10.

[41] Scheer, "Are Emotions a Kind of Practice?", 204. My italics.

[42] Fay Bound Alberti, "Introduction: Emotion Theory and Medical History". In *Medicine, Emotion and Disease 1700–1950*, ed. Fay Bound Alberti (Palgrave Macmillan, 2006), xviii.

The apparently contradictory meeting of spontaneity and convention is, I will argue, crucial to the energy and efficacy of self-wounding in this period. Acts of self-wounding were represented as meaningful and useful to those who practised them, but the manner of their meaning or use depended on the chronological, generic and social contexts in which they were presented. When it appeared in Elizabethan drama, for example, self-wounding served characters as a means of expression which claimed to be more immediate and truthful than verbal rhetoric, and thus aimed to secure honour and trustworthiness (or, in Bourdieu's terms, symbolic and social capital) for the practitioner. In other contexts, self-wounding was a form of what Scheer calls 'emotion-as-practice', a way of *doing* feeling which drew on prevailing beliefs about the rational imperative to manage one's humours and passions (discussed in more detail in Chap. 4). Still elsewhere, the way in which self-wounding mimicked forms of judicial or social control enabled it to act as a form of resistance, undermining or exaggerating the subjection of individuals to normative codes. The acts this study examines are always interactionist, always embedded in social, cultural and political contexts. We cannot recover how it 'really' felt for individuals to injure themselves, but we can see how the meaning of such acts was created, changed or denied in the dynamic textual environment of early modern England.

1.5 Outline

The following two chapters of this book show how self-wounding was represented in the late sixteenth and early seventeenth century as a communicative practice, which leant on the somatic as a powerful, if unstable, basis for asserting the sincerity of one's convictions. In Chaps. 4, 5 and 6, I consider how people used self-injury to negotiate their own sense of self, to manage their emotions or resist existential threat. Chapter 7 is a conclusion of sorts, a brief story of eighteenth-century self-injury through which I question why and when beliefs about this practice changed. Chapter 8 serves as 'coda', reflecting on wider questions of how we relate to historical phenomena and detailing the co-creative projects related to this work.

In Chap. 2, I examine representations of self-stabbing and self-cutting in two canonical plays, Shakespeare's *Julius Caesar* and Marlowe's *Tamburlaine the Great, part 2.* Both plays, I argue, situated self-injury in relation to discourses about rhetoric and in particular gesture, a practice which both expresses thought and helps to facilitate thinking. Portia and

Tamburlaine use self-injury as a persuasive strategy, and they present the act of self-wounding, and the wound itself, as evidence of sincerity which exceeds that which can be offered by verbal rhetoric alone. Self-injury supposedly reveals the practitioner's true 'inside' rather than their exterior show. At length, however, this supposition relied on a delicate balance between the symbolic potential of self-injury and its material effects on the body. In *Julius Caesar*, Portia's 'voluntary wound' upholds her claims to constancy and provides her with a means to be taken seriously in a male-dominated oratorical culture. However, self-wounding was not only formulaic, for its status as pain-causing was what set it apart from other kinds of communicative act. This, I argue, is what causes Tamburlaine's self-inflicted cut, which likewise aims at impressing an audience, to fall short of his heroic intent, becoming a 'pitiful sight'. Though Tamburlaine couches his self-wounding in grand rhetoric, he makes the mistake of denying that it hurts him, and in so doing he robs it of significance.

Chapter 3 extends this theme of self-injury as communication to consider how self-injury's persuasive potential interacted with contemporary discourses about one's rights over and responsibilities towards the body. Discussions of suicide show that while self-wounding was generally considered as a separate issue, it sometimes became freighted with some of the moral and religious stigma which attended self-killing, as well as with the countervailing neo-stoic ideology in which total dominion over the body was a praiseworthy aim. The latter ideology is played out in Thomas Kyd's popular 1587 play *The Spanish Tragedy*, wherein the protagonist bites out his tongue as part of a spectacular revenge plot. I argue that while this episode may seem nonsensical, it can be understood within the context of Stoic stories in which autoglossotomy functions as an act of 'necroresistance'. Tongue-biting allows Hieronimo to cast himself as hero, and in so doing to interpellate his captors as tyrants. Nonetheless, moral ambiguity about his actions remains, and later revisions of *The Spanish Tragedy* raise the possibility that Hieronimo's actions are not those of a freedom fighter, but of a madman, controlled by metaphysical forces beyond his understanding.

I turn in the next section to exploring how self-injury affected the practitioner's own social and emotional experiences. In Chap. 4, I examine representations of violent lament, in which grief-stricken individuals—usually women—incorporate face-scratching, hair-pulling and breast-beating into ritualistic expressions of sorrow. Inherited from the classical tradition, lament persisted in the early modern period as a literary form. Yet, it sat uneasily with contemporary thinking about the proper governance of the emotions, which sought to understand how the social, physical and spiritual aspects of the emotional 'ecosystem' could be balanced with one another. On the one hand, imperatives against 'immoderate' sorrow depicted prolonged grief as both unchristian and potentially detrimental to the griever, who might whip up the melancholy humours into such abundance that they became dangerously ill. On the other hand, early modern audiences continued to respond to imaginative texts which depicted extremes of feeling, partly because they were felt to offer a cathartic release of emotions. In Edmund Spenser's *The Faerie Queene* and in various of Shakespeare's plays and poems (including *The Rape of Lucrece*), representations of violent lament provide an opportunity to think through the puzzle of early modern emotional expression, and whether 'doing sadness' was helpful or harmful.

Chapter 5 also looks at the way in which self-wounding might shape social, emotional and physical experiences, though in a radically different form. It explores the surprising stories of men who cut off their own testicles, as they are relayed in medical texts, comical ballads and books of curiosities. Unlike the protagonists of many other accounts in this book, such men were anything but heroic. Their actions fit into a culture in which to lose one's testicles was an occasion for shame, and a good reason for social ostracisation. Investigating why men would choose this course of action, I argue that self-castration was often represented as a reaction against the pressures of contemporary masculine ideals. Men reportedly gelded themselves because they suspected their wives of being unfaithful or because of impotence, poverty or sexual shame. In this way they repudiated the system which equated masculinity with sexual prowess and the ability to dominate one's household. Yet men's self-castration also shows them to be drawn to shame and ignominy. In the latter part of the chapter, I consider how far we might view self-gelding as enacting a kind of 'social masochism' or even social suicide, which deliberately aims at the annihilation of the person involved as a gendered social subject.

Chapter 6 brings together the preceding themes of self-determination and self-annihilation. I examine the horrifying story of Saint Ebba, an eighth-century nun who, together with her convent sisters, apparently cut off her own nose in order to avoid rape by Viking marauders. Ebba's story is something of a puzzle. It was, as I demonstrate, surprisingly popular among seventeenth-century readers. Yet it was also difficult to place—Ebba's actions could be viewed as morally problematic, and they bore an uncomfortably close relationship to physical mortification, a practice decried by most reformed Christians. In analysing the unlikely durability of this story, I speculate that in the seventeenth century, it was revitalised partly as a result of other discourses which cast women's alteration of their bodies as empowering. Tales of female self-starvation in this period emphasised the potential of such acts as assertions of individual spiritual and economic agency. Meanwhile, stories of Amazon breast removal cast the self-alteration of that tribe as modification rather than mutilation, a practical step which also cemented belonging in a radically different sort of society. Ebba's story was not emancipatory, however. Ultimately, the enraged attackers took their revenge by torching the still-occupied convent. I argue that this outcome spoke to an early modern climate in which self-injured women, like rape victims, had no imaginable place within the heteronormative economy, and therefore 'had' to die.

Chapter 7, the conclusion, looks forward to eighteenth-century perceptions of self-injury. In this century, I argue, representations of self-harm became less concerned with the agency and needs of the self-injuring person. Through an extraordinary story of self-stabbing by a prison inmate, I suggest that people who self-injured no longer had their injuries 'read' on their own terms, but were instead interpreted according to the agenda of their audience. Serious injuries became more likely to be viewed as unsuccessful attempts at suicide, while minor ones were dismissed as attempts at manipulation. I suggest some possible reasons for this shift, and highlight the factors which made the sixteenth and seventeenth centuries a unique period in the history of self-wounding.

The final chapter of this book departs from a traditional literary-historical format in order to think more expansively about what might be gained, or endangered, in bringing historical narratives of self-wounding into dialogue with twentieth- and twenty-first-century thinking on non-suicidal self-injury. As I show, the general trend in public and institutional attitudes towards self-wounding has been one of medicalisation. Non-suicidal self-injury entered the Diagnostic and Statistical Manual of Mental Disorders as a 'condition for further study' in 2013, meaning that more research is needed before the phenomenon can be officially termed a 'disorder'. However, in recent years, scholars including Baptiste Brossard, Peter Steggals and others have suggested an alternative interpretation of this practice based on sociological observation—something people *do*, rather than an affliction which they *have*.[43] This aligns more closely with the multiple interpretive frameworks which this book argues were available for understanding self-injury in the early modern period; though as I show, the two should not be conflated. As well as contemporary scholarship, this chapter is informed by personal reflection and creative practice, particularly art and theatre projects run from 2021–2024. The insights generated by those projects offer another perspective on the problem of 'presentism', fuelled by the honesty and generosity of the participants. Finally, I draw attention to the potential pitfalls of public engagement versus co-creation, and the dangers of seeking to 'impact' non-academic participants.

[43] Peter Steggals, "Making Sense of Self-Harm: Exploring the Cultural Meaning and Social Context of Non-Suicidal Self-Injury". PhD diss., Newcastle University, 2013. http://theses.ncl.ac.uk/jspui/handle/10443/2143; Peter Steggals, Steph Lawler, and Ruth Graham, "The Social Life of Self-Injury: Exploring the Communicative Dimension of a Very Personal Practice," *Sociology of Health & Illness* 42, no. 1 (2020): 157–70, https://doi.org/10.1111/1467-9566.12994; Baptiste Brossard, *Why Do We Hurt Ourselves?: Understanding Self-Harm in Social Life* (Indiana University Press, 2018); Amy Chandler, "Self-Injury as Embodied Emotion Work: Managing Rationality, Emotions and Bodies," *Sociology* 46, no. 3 (2012): 442–57, https://doi.org/10.1177/0038038511422589; Gary Taylor, *Castration: An Abbreviated History of Western Manhood* (Routledge Ltd, 2000); Millard, *A History of Self-Harm in Britain*; Sarah Chaney, "Self-Control, Selfishness and Mutilation: How 'Medical' Is Self-Injury Anyway?," *Medical History* 55, no. 3 (2011): 375–82.

CHAPTER 2

Persuasion: Self-Wounding as 'strong proof' in *Julius Caesar* and *Tamburlaine the Great, Part Two*

PORTIA: *I was smarter than they believed, smarter than my husband believed. You see, I understood the political climate and my husband's role in it. Increasingly, I became aware of his involvement in the conspiracy against Caesar. I also saw the doubts he was having about it. But, still, he would not confide in me. Still, he would see me as just a woman, someone to just be quiet and submit to him. So, I spoke in the only language these brave men seemed to understand. I took a sharp knife to my thigh and let that do the talking.*[1]

From the middle of the twentieth century onward, self-injurious behaviours have been described by some medical professionals as 'suicide gestures'. The defining features of these 'gestures' include dramatic presentation, 'low medical lethality' and 'low or no intent to die'.[2] The suicide gesture is instead variously described as a bid for attention, an attempt to manipulate others, or, slightly less judgementally, a communication of distress. These descriptions are, as recent scholarship has recognised, problematic, and they reflect the negative attitudes people who

[1] Quote from *I Am Worthy*, play devised by participants in the Self Injury Through Time theatre group, 2024.

[2] Nicole Heilbron, Jill Compton, Stephanie Daniel, and David Goldston. "The Problematic Label of Suicide Gesture". *Professional Psychology: Research and Practice* 41, no. 3 (2010): 221–27, (222–24); Andrew M. Huckins-Noss, Kathryn R. Fox, Alexander J. Millner, Matthew K. Nock, and Jessica Stubbing. "The Classification of Suicide Gestures: Examining the Validity, Measurement, and Level of Intent of a Controversial Construct". *Suicide and Life-Threatening Behaviour* 53, no. 2 (2023): 188–97.

A. Skuse, *Hurt Feelings*, Palgrave Studies in Literature, Science and Medicine, https://doi.org/10.1007/978-3-032-04565-2_2

self-injure may experience when they seek help in primary care or accident and emergency settings. Moreover, the word 'gesture' may imply that self-injury undertaken for any purpose other than suicide is insincere, illegitimate or less meaningful than 'genuine' suicide attempts.

The terminology of 'suicide gestures' therefore contrasts strikingly with early modern discourses, in which the gestural quality of self-wounding was understood as a part of how the practice *made* meaning and attempted—albeit with difficulty—to convey sincerity. Throughout this book, I will show that early modern self-wounding functioned as a practice which might accomplish a purpose or send a message, yet which often also resisted interpretation, retaining in the end a certain degree of unknowability even when it seemed to be thoroughly explicated. In this chapter, I focus on self-wounding as a communicative act, embedded in Renaissance theories about rhetoric, gesture and cognition.

The first part of the chapter examines early modern writings about the use of the body in rhetoric, and particularly gesture as a communicative tool akin to language. I show how early modern theories of gesture share features with modern cognitive psychological research on the subject. Both discourses emphasise—though in very different ways—the status of gesture as both spontaneous and formulaic, a 'natural' expression shaped by cultural rules. Moreover, both find that gesture may materially affect both the gesturer and their audience. Self-wounding, as it is represented by early modern authors, shares many of these gestural features. However, self-wounding may also go beyond gesture in its ability to create tangible effects in the form of bodily pain and scars, so that both the temporal act and its longer-lasting trace constitute 'strong proofs'. In the second and third parts of the chapter, I apply these insights to two literary examples of self-wounding—Portia's 'voluntary wound' in *Julius Caesar*, and Tamburlaine's self-inflicted cut in *Tamburlaine the Great, Part II.* Both these texts show how self-wounding is used to construct identity by presenting the protagonist to their audience as knowable 'through and through', and thus allaying anxieties about the gap between presented and authentic selves. However, where Portia's self-stabbing broadly succeeds in accomplishing her aims, Tamburlaine's falls short, because it relies on a self-conception that seeks to disavow intersubjectivity in favour of absolute self-reliance. In both works, self-injury poses questions about how thoroughly it is possible to access the experience of another person.

2.1 'Our swords shall play the orators': Self-Wounding, Gesture and Emotion

In a 2012 article on 'gestural sense-making', the philosopher Elena Cuffari argues eloquently for the importance of gesture as a foundational aspect of human communication:

> The ubiquitous human practice of spontaneously gesturing while speaking demonstrates the embodiment, embeddedness, and sociality of cognition. The present essay takes gestural practice to be a paradigmatic and illuminating example of a more general claim: human cognition is social insofar as our embedded, intelligent, and interacting bodies select and construct meaning in a way that is intersubjectively constrained and defeasible.[3]

Cuffari's assertions are part of a renewed interest in the uses of gesture which has seen scientists and phenomenologists agree that gesture is a form of communication akin to speech. Indeed, it is frequently characterised *as* a language, made meaningful 'by virtue of some of the same conditions and constraints that render verbal language use meaningful', and 'something that knows of success or failure'.[4] This bold claim is followed by an even bolder one: gesture is imbricated with cognition itself, such that as Andy Clark argues,

> the physical act of gesturing is part and parcel of a coupled neural-bodily unfolding that is itself usefully seen as an organismically extended process of thought. In gesture, we plausibly confront a cognitive process whose implementation involved machinery that loops out beyond the purely neural realm.[5]

Rather than merely being employed to convey a fully formed concept, gesture is, as Merleau-Ponty argues, integral to the 'accomplishment of thought'.[6] It is both communicative and constitutive, woven through the process of forming and expressing ideas. One explanation for how this

[3] Elena Cuffari, "Gestural Sense-Making: Hand Gestures as Intersubjective Linguistic Enactments". *Phenomenology and the Cognitive Sciences* 11, no. 4 (2012): 599–622. https://link.springer.com/article/10.1007/s11097-011-9244-9, 600.

[4] Cuffari, 'Gestural Sense-Making', 604, 605.

[5] Andy Clark, "Gesture as Thought?", in *The Hand, an Organ of the Mind: What the Manual Tells the Mental*, ed. Zdravko Radman (MIT Press, 2013), 257.

[6] Clark, "Gesture as Thought?", 260.

might work involves 'mirror neurons' which 'fire both when an animal performs some intentional action and when it sees another animal performing the same action'.[7] Some neuroscientists suggest that our *own* gestures might cause our mirror neurons to react and thus 'activate' the part of the brain that responds to intentional actions.[8]

Such formulations entwine gesture with cognition, and imagine cognition itself as embodied and social. Moreover, whilst the discussion of embodied or extended cognition most obviously reaches forward, towards transhumanism, it also reaches *back*. Inadvertently, scholars in cognitive psychology have found themselves on common ground with early modern thinkers for whom it seemed self-evident that thoughts and emotions were the product of psycho-physiological, spiritual and external factors. Miranda Anderson, whose book *The Renaissance Extended Mind* addresses this proximity, argues that 'the Renaissance notion of reading another mind through their gestures and physiognomy is linked to capacities now understood as enabled in part through mirror neurons'.[9] As a means of 'mind-reading', gesture was afforded substantial social and cultural importance. Adam Kendon describes how the sixteenth century saw increasing interest in the communicative potential of the body driven by diverse factors. Voyages into the New World, he argues, led to speculation on the potential of gesture to form a 'universal language', while at home, a new emphasis on the pedagogical role of clergy (both Catholic and Protestant) sparked discussion about the best ways to absorb and persuade one's audience.[10] Furthermore, conduct texts aimed at cultivating the polished gentleman or courtier dwelt on bodily comportment as an essential aspect of this self-fashioning.[11] Unsurprisingly, therefore, gesture acquired considerable significance in early modern textbooks on rhetoric. In his 1588 *The Arkadian Rhetorique*, for instance, Abraham Fraunce attested that gesture was 'both more excellent and more universall than voyce: as belonging not onelie to those that use the same speech, but generallie to all people, yea

[7] Clark, "Gesture as Thought?", 259. It is worth noting that mirror neurons, though much talked-about, remain a subject of debate. It is unclear if mirror neurons are a separate class of cells, and if so, which species have mirror neurons and how far they can be said to be responsible for qualities such as empathy.

[8] Clark, "Gesture as Thought?", 259.

[9] Miranda Anderson, *The Renaissance Extended Mind* (Springer, 2015), 158–59.

[10] Adam Kendon, *Gesture: Visible Action as Utterance* (Cambridge University Press, 2004), 21.

[11] Kendon, *Gesture*, 22.

to beasts and senseless creatures, as the verie pictures which being dumme, yet speake by gesture an action'.[12] Francis Bacon likewise argued that 'the Motions of the countenance and parts … disclose the present humour and state of the mind & will'.[13] Such arguments drew from the writings of classical rhetoricians such as Quintilian, whose *Institutio Oratoria* had been rediscovered in the fifteenth century. In a brief but influential passage, the *Oratoria* gave detailed instructions for the use of hand gestures and argued that 'the hands themselves virtually speak … is it not the case that we use them to demand, promise, summon, dismiss, threaten, entreat, show aversion and fear, question and deny?'[14]

As well as the importance of gesture, it is the *way* in which gestures were supposed to work which links early modern and modern. This is evident from the most comprehensive seventeenth-century works on gesture, John Bulwer's *Chironomia, or, the Art of Manuall Rhetorique,* and *Chirologia, or, the Natural Language of the Hand* (both 1644). Not only did Bulwer provide an encyclopaedic list of gestures in these works, but publication of the two texts together emphasised that the formal aspects of gesture explored in *Chironomia* and the natural expressiveness described in *Chirologia* were opposite sides of the same coin:

> [T]he first Artificers of Manuall Rhetoricke, hit on the right veine of Oratorie, when … they tooke in hand that polite device, and elegant design of reducing the usuall gestures of Nature into strict rules of Art, preparing the undigested motions of Nature, and making them more formall, and fit for the intention of Rhetoricke, whose life and force they made much to consist in the just demeanour of the Hand, whose motions appear as emphaticall to the eye, as speech doth to the eare.[15]

Much as the stomach digested food into nutritious chyle, commonly used gestures adopted the 'natural' impulses of the human body and made them formulaic, hence more easily readable. This was not the same as

[12] Abraham Fraunce, *The Arkadian Rhetorike: or The praecepts of rhetorike made plaine by examples* (Thomas Orwin, 1588), fol. J4v.

[13] Francis Bacon 2000b, 94 quoted in Anderson, *The Renaissance Extended Mind*, 158.

[14] Quintillian, *Institutio Oratoria* 11.3.85–86 translated by and quoted in Jon Hall, "Cicero and Quintilian on the Oratorical Use of Hand Gestures". *The Classical Quarterly* 54, no. 1 (2004): 143–60. https://www.jstor.org/stable/3556289, 150.

[15] John Bulwer, *Chironomia, or, the Art of Manuall Rhetorique* (Tho(mas) Harper, 1644), 16–17.

saying that the gestures used by orators were faked copies of natural gestures; on the contrary, they still contained genuine emotion. As Joseph Roach elaborates, 'What orators and stage players do … is to discover the passions of the mind with their bodies—larynx, limbs, torso, and head together—thereby transforming invisible impulse into spectacle and unspoken feeling into eloquence'.[16] Lynn Tribble likens the repeated practising of gestures advocated by Bulwer to the practice of athletes, sedimenting a movement in the body so that it can be produced effortlessly at the appropriate moment.[17] Thus, for example, Bulwer described the striking of the breast as a gesture both 'very patheticall … in Nature, & Rhetorical in Art'. The force with which one struck the breast was a matter of both judgement and personal emotion:

> [this action] they doe, for the most part, to testifie anguish of minde, repentance, and matters of Mortification … Which Rhetoricall action of the Hand is not alwaies (to an inch) framed by the receipts of Rhetoricians, nor by line and levell fitted to the rule of Art, nor weighed, as 'twere, in the goldsmiths balance; for they who assume this gesture, strike their breast with an audible stroake, when they judge it fit for their purpose; although some, who are more studious of Eloquence, doe not heartily admit of this loud contact of the Hand.[18]

This dual status, artful and heartfelt, was crucial to understandings of gesture. Actions such as breast-beating acquired the status of tropes because they were natural, and they were understood as natural because they conformed to a commonly understood 'language'. John Wesley explains how

> gestural expressions of intention and emotion comprised action we might not consider formal or formulaic, but which, for early modern actors, orators, and spectators, was expedient, mimetic, and a source of bodily 'puns.' Foundational to this way of conceptualizing representation is the idea that nature—in the form of its visceral liquid passion—really does compel the body to behave involuntarily in specific, recognizable ways … For early moderns, the emotions had 'intrinsic physical manifestations', but, more

[16] Joseph R. Roach, *The Player's Passion: Studies in the Science of Acting* (University of Delaware Press, 1985), 32–33.

[17] Evelyn Tribble, *Cognition in the Globe: Attention and Memory in Shakespeare's Theatre* (Palgrave Macmillan US, 2011), 34.

[18] Bulwer, *Chronomia*, 46–47.

> than just the signs of internal operations, it was felt that the adoption of these physical forms could trigger their assigned emotions, an outer-to inner causal relationship that anticipates current thinking on the role of gesture in human cognition and language.[19]

As Wesley, Tribble and Anderson all point out, this duality aligns early modern theories with the mirror neuron theory described above, and more generally with recent cognitive psychology which suggests that enacting a gesture may bring about its associated emotion (smiling making one feel happy, for instance).[20] Clark, for example, suggests that 'by materializing imagistic thought in physical gesture we create a stable physical presence that may productively affect and constrain the neural elements of thought and reason'.[21] Gesture, he argues, is somewhat akin to thinking out loud, or thinking through writing; not something overlaid onto fully formed ideas but rather a constituent thereof.

Early modern people too imagined gesture as both expressive *and productive* of emotion, a belief based in their understanding of oratory.[22] To fulfil this function, however, it had to seem genuine, and this was most convincingly achieved if—in some sense at least—it *was* genuine. It was a truism of texts on rhetoric that the effective orator needed to embody the emotion he wished to inculcate in his audience. In his *The Passions of the Minde in Generall*, Thomas Wright insisted that

> Cicero expresly teacheth that it is almost impossible for an oratour to stirre up a passion in his auditors, except he be first affected with the same passion himselfe ... It cannot be that he which heareth should sorrow, hate, envie, or feare any thing, that he should be induced to compassion or weeping, except all those motions the oratour would stirre up in the judge, be first imprinted and marked in the oratour himselfe.[23]

[19] John Wesley, "Original Gesture: Hand Eloquence on the Early Modern Stage," *Shakespeare Bulletin* 35, no. 1 (2017): 65–96, https://doi.org/10.1353/shb.2017.0003, 67; see also Farah Karim Cooper, *The Hand on the Shakespearean Stage: Gesture, Touch and the Spectacle of Dismemberment* (Bloomsbury Publishing, 2016), 146.

[20] Tribble, *Cognition in the Globe*, 31–33.

[21] Clark, 'Gesture as Thought?', 259.

[22] Cooper, *The Hand on the Shakespearean Stage*, 146. On the similarities and differences between modern and early modern models of extended or embodied cognition, see Anderson, *The Renaissance Extended Mind*.

[23] Thomas Wright, *The Passions of the Minde in Generall. Corrected, Enlarged, and with Sundry New Discourses Augmented*. (Valentine Simmes and Adam Islip for Walter Burre and Thomas Thorpe, 1604), 172.

To do so, Quintilian advised that the orator stirred up *phantasiai*, imagining sorrowful, frightening or joyful events in such detail that they felt and expressed those emotions in every aspect of their voice and gesture.[24] His instructions for stirring emotion reflected the contention in Aristotle's *De Anima* that images conjured by the imagination (*imago animi*) moved the soul in the same way as objects actually present to the sight.[25] Moreover, this logic extended to the audience, who experienced the represented emotions in profound terms. Truly effective rhetoric, whether manual or verbal, could literally stir the spirits of those who saw or heard it, since according to the dominant humoral model, the passions (motions of the sensitive soul provoked by thoughts and sensations) travelled to the heart and caused that organ to draw to itself the humours most apt to a certain emotion. These in turn produced the physiological sensations which were an important part of that emotion, the coldness of fear or the heaviness of grief.[26]

As these effects suggest, the considerable power of rhetoric and gesture came with ethical problems. The suggestion that orators should 'personate' or embody the emotion they wished to inculcate blurred distinctions between what emotions truly belonged to oneself and which were assumed. Writing on the early modern grammar school system in which oratory and accompanying gesture was learned, Lynn Enterline notes that exercises in prosopopoeia—imagining the speech of 'weeping Hecuba', for instance—offered schoolboys the chance to express their own negative feelings: 'to turn a figure that weeps for others into a way of weeping for oneself and one's own cause for complaint'.[27] This raised questions about the nature of *ethos*, or character. If one's own feelings were best approached through exemplars, that suggested that 'for speaking subjects, relationships to others are constitutive'; that is, that character was not merely expressed through but actually formed via the performance of ethos.[28] As Davis

[24] Robert Cockcroft, *Rhetorical Affect in Early Modern Writing: Renaissance Passions Reconsidered*. (Palgrave Macmillan UK, 2003), 52.

[25] Roach, *The Player's Passion*, 25.

[26] Katharine A. Craik and Tanya Pollard, *Shakespearean Sensations: Experiencing Literature in Early Modern England* (Cambridge University Press, 2013), 11. See also Indira Ghose, *Shakespeare and Laughter: A Cultural History* (Manchester University Press, 2008).

[27] Lynn Enterline, *Shakespeare's Schoolroom: Rhetoric, Discipline, Emotion* (University of Pennsylvania Press, 2011), 131–32.

[28] Enterline, *Shakespeare's Schoolroom*, 133.

identifies, this implied a worryingly labile kind of selfhood: 'The equivocal process of ethopoetic rhetoric, as the making and unmaking of individual and social character, becomes of prominent concern'.[29]

This lability became especially troubling in relation to the stage. One obvious problem was that acting could be considered lying, leading to moral degeneracy among actors.[30] But potentially even worse was that plays excited dangerously strong emotions in their audience. Heywood's *Apology for Actors* boasted that 'so bewitching a thing is lively and well spirited action, that it hath power to new mold the hearts of the spectators and fashion them to the shape of any noble and notable attempt'.[31] The noble and notable, however, was not the only direction in which an audience could be moved. Gesture was central to theatre's capacity to 'infect [the audience's] minde with wicked passions, so that in their secret conclaves they play the Sodomite, or worse'.[32] Anthony Munday complained that 'It is marvelous to consider how the *gesturing* of a plaier, which Tullie termeth the eloquence of the bodie, is of force to move, and prepare a man to that which is il'.[33] Both apologies for and attacks on the theatre recognised how the system of passions and humours which knit imagination, emotion and physiology also allowed for slippage between characters, actors and audiences.

* * *

How were these possibilities and anxieties played out in relation to self-wounding? The question is complicated, because there was no clear line between clearly gestural but potentially hurtful actions such as breast-beating, as described by Bulwer, and the more extreme territory of self-stabbing, cutting or even auto-amputation. Nonetheless, the gestural *quality* of such events is one of the threads that unites them, even as their effects vary widely. On stage at least, these practices, as I demonstrate

[29] Lloyd Davis, *Guise and Disguise: Rhetoric and Characterization in the English Renaissance* (University of Toronto Press, 1993), 35.

[30] Stephen Gosson, *Playes Confuted in Five Actions* (Thomas Gosson, 1582), fol. E5r; see also John Greene, *A Refutation of the Apology for Actors* (W. White, 1615), 55.

[31] Thomas Heywood, *An Apology for Actors* (Nicholas Okes, 1612), fol. B4r.

[32] Greene, *A Refutation of the Apology for Actors*, 51.

[33] Anthony Munday, *A Second and Third Blast of Retrait from Plaies and Theaters* (Henrie Denham, 1580), 95. My italics; see also Philip Stubbes, *The Anatomie of Abuses* (New Shakspere Society, 1877 (orig. 1583)), x.

below and in the following chapter, took place for explicitly communicative and persuasive purposes—they too were forms of 'manual rhetoric' which put the body to use. Indeed, this is a principal difference between early modern self-wounding and modern categories of non-suicidal self-injury (NSSI). In contrast to modern discourses about NSSI, which often show it to be secretive and stigmatised, early modern self-wounding is overwhelmingly depicted as something which admits of discussion and examination (even when, like Portia's wounding, it is not explicitly public).[34] Like gesture, self-wounding also involved cultural scripts; as this book shows, when early modern people read of a man gelding himself, or watched a character bite out their tongue on stage, they might see those acts as having something in common with the self-castration of Origen or the autoglossotomy of Zeno the Stoic. Self-wounding thus relied for its intelligibility on the same mixture of codification and natural spontaneity that Bulwer posited as the basis for successful gestures. It is for this reason that people who self-wounded, even in real life, were not generally identified as mad. Instead, their actions were usually placed in social or personal context; self-wounding, it was implied, was an idiom that could be understood, provided one had the correct interpretive information.

At the same time, however, practices of self-injury exceeded gesture in several respects. First, self-wounding was enduring where gesture was ephemeral (though clearly, the 'endurance' of an on-stage wound depended on the audience's suspension of disbelief). Not all kinds of wounding left a scar or a permanent bodily difference, but most had some physical trace which remained visible hours, days or months after the event. As I will argue of Portia's voluntary wound, this extends the time in which the self-injury is readable. In so doing it allows the self-inflicted wound to be 'staged' in a different context from the one in which it is inflicted, and it facilitates explanation and reinterpretation of the wound and its meanings. In this respect, self-wounding is to gesture as writing is to speech; it inscribes the ephemeral event, investing it with significance and setting it as a point of reference. One could dissimulate the nature or potency of one's thoughts or emotions, but wounding oneself provided surety, as far as humanly possible, that one really was experiencing pain. As

[34] Of course, there may have been another private world of early modern self-wounding to which we are not privy because it has not been recorded. In the texts I have examined, however, there is little suggestion that this is the case. Self-wounding is rarely concealed and is often undertaken in front of witnesses.

a material form of expression, it brought into being that which it sought to express, a feat which depended for its intelligibility on the translation of emotional pain into physical terms. This accords with the contention of Jan Frans van Dijkhuizen and Karl Enenkel that early modern people 'locate[d] the meaning of pain first and foremost in its overwhelmingly bodily nature', and so sought to translate emotional into physical sensations.[35] Whereas the modern study of pain often focuses on its cultural aspects, they argue, '[e]arly modern perceptions of pain frequently work in precisely the opposite direction: they invoke the physicality of pain to invest other, non-bodily categories of experience with the authority and palpable reality of bodily sensation'.[36] There were certainly problems with a translation from rhetorical to physical; as the analyses below demonstrate, self-wounding is a communicative practice which lacks the specificity of gesture, and therefore relies on careful situating to be 'correctly' interpreted. Nonetheless, for visceral affective potential, self-directed violence matches and outdoes gesture, with the onlooker identifying with the protagonist even against their will. There is little to match the involuntary wince induced by the sight of another person being injured.

In the case studies below, I analyse how the communicative potential of self-wounding as something akin to gesture is deployed in two popular plays of the early modern period, Shakespeare's *Julius Caesar* (1599) and Marlowe's *Tamburlaine the Great, Part II* (1592). I argue that in both cases, the protagonists (Portia and Tamburlaine) seek to utilise the 'wince effect' to engage their on- and off-stage audiences and compel attention to the messages they attach to their wounds. In so doing, they highlight both the potential and the pitfalls of self-wounding as persuasive practice.

2.2 *Julius Caesar* and Portia's 'Voluntary Wound'

> I grant I am a woman, but withal
> A woman that Lord Brutus took to wife.
> I grant I am a woman, but withal,
> A woman well reputed, Cato's daughter.

[35] Jan Frans van Dijkhuizen and Karl A. E. Enenkel, "Introduction: Constructions of Physical Pain in Early Modern Culture," in *The Sense of Suffering: Constructions of Physical Pain in Early Modern Culture* ed. Jan Frans van Dijkhuizen and Karl A. E. Enenkel (Brill, 2008), 4–5.

[36] Dijkhuizen and Enenkel, "Introduction", 6.

> Think you I am no stronger than my sex,
> Being so fathered and so husbanded?
> 'Tell me your counsels; I will not disclose' em'.
> I have made strong proof of my constancy,
> Giving myself a voluntary wound
> Here in the thigh. Can I bear that with patience,
> And not my husband's secrets? (2.1.291–301)

In the plethora of stabbings which populate Shakespeare's *Julius Caesar*, Portia's voluntary wound is often read as one more instance of the 'histrionic appeal' of onstage violence.[37] As Marshall observes, this stabbing works alongside Calphurnia's dream of Caesar's statue spouting blood as part of an 'image cluster … around the central motif of Caesar's body pierced by wounds'.[38] As the play's first violent act, it also prefigures the voluntary deaths—including Portia's own—which shape the drama's ending, and the eventual fate of the Roman empire. Here, however, I wish to argue that the status of Portia's wound as voluntary sets it apart from both the interpersonal conflict which defines Caesar's death and the cascade of suicides with which the play concludes. Portia's act stands out as a testament to the potential of self-wounding as a persuasive tool, and a case study in the relationship between a wound and the gesture it represents.

Portia's stab has an explicit rhetorical aim. By stabbing herself, she wishes to construct an *ethos* which will persuade her audience (that is, Brutus) that she is a worthy recipient of his secret. Specifically, she aims to prove her constancy, a quality that is prized throughout the drama from Caesar's self-construction as 'northern star' to Brutus's resolute attachment to a nebulous ideal of 'honour'. The play's emphasis on constancy is timely, since as I discuss in the following chapter, Stoicism became an exciting tool for thought in the early modern period. *Julius Caesar* was published only two years after the English translation of Justus Lipsius's *De Constantia*, a neo-Stoic treatise which dwelt on the role of Stoicism in times of political turbulence. However, it is clear from the outset that this playworld is one in which the appearance of constancy matters as much, or more, than its actuality. Analysing *Julius Caesar's* construction of

[37] Maik Goth, "'Killing, Hewing, Stabbing, Dagger-Drawing, Fighting, Butchery': Skin Penetration in Renaissance Tragedy and Its Bearing on Dramatic Theory". *Comparative Drama* 46, no. 2 (2012): 139–62, (140).

[38] Cynthia Marshall, "Portia's Wound, Calphurnia's Dream: Reading Character in Julius Caesar". *English Literary Renaissance* 24, no. 2 (1994): 471–88 (473).

republicanism and masculinity, Coppelia Kahn notes how the drama's male characters 'present themselves as actors both histrionic and historical amid the "lofty scene" of politics, performing their parts with the "formal constancy" of stage actors before an audience of posterity'.[39] Brutus, Caesar and Cassius all, at points, speak about themselves in the third person, and employ the stately style of public oratory even during private conversations. Personal subjectivity is constructed almost exclusively in relation to political roles, or as Kahn puts it, 'constituted in and by a social field'.[40]

The primacy of rhetoric in *Julius Caesar's* playscape is emphasised early and often, beginning in 1.2, in which Cassius attempts to persuade Brutus of Caesar's monarchical ambitions. Conducting their ostensibly private conversation on the edge of a large crowd, Cassius presents his own ethos as evidence for his good intentions:

> Were I a common laugher, or did use
> To stale with ordinary oaths my love
> To every new protester; if you know
> That I do fawn on men and hug them hard
> And after scandal them, or if you know
> That I profess myself in banqueting
> To all the rout, then hold me dangerous. (1.2.160–66)

Cassius relies on his body to prove his authenticity; he asserts that he cannot fake the reaction of laughing, nor the gesture of hugging, and thus these things can be read in him as natural impulses. He is, he argues, just what he appears to be, whereas Caesar's apparent majesty conceals an effeminate weakness and 'feeble temper' which is likewise revealed by his body—specifically, his lack of physical prowess during the swimming match (1.2.219). Moreover, Cassius tests his rhetorical capabilities further when he presents Brutus with a version of himself (that is, of Brutus) in which the worlds of public and private are enmeshed so closely that his innermost thoughts tend only towards the republicanism which defines his patrician status. Drawing on an Aristotelian model of character as formed by and represented through actions, Cassius promises that 'I, your glass, / Will modestly discover to yourself / That of yourself which you

[39] Coppélia Kahn, *Roman Shakespeare: Warriors, Wounds and Women* (Taylor & Francis Group, 1997), 78.

[40] Kahn, *Roman Shakespeare*, 79.

yet know not of' (1.2). As Xinyao Xiao observes, this speech elides the gap between 'outward favour' and 'virtue … in you', such that 'Honor then becomes simultaneously Brutus's observable, performative exterior, and his inexpressible, substansive core'.[41] Looping between 'you', 'yourself' and 'I', between knowing and discovering, Cassius also slips between ethopopoeia and prosopopoeia, and both tend to elide social appearance with personal interiority. This tendency extends throughout the play, usually escaping the notice of those whom it affects. Noting the tautology of a system in which virtue is a 'Roman' attribute and Roman-ness itself is a virtue, Miles Geoffrey finds that despite the play's emphasis on constancy, 'the truth is constituted by judgements arrived at through a process of public observation, discussion, and persuasion'.[42] Nowhere is this more apparent than in the exchange between Brutus and Ligarius, who knocks on the door and interrupts Portia's reconciliation with her husband. Initially presenting as 'a sick man' with 'a feeble tongue' (2.1.309, 312), Ligarius swiftly declares 'I am not sick if Brutus have in hand / Any exploit worth the name of honour … I shall discard my sickness' (2.1.315–16, 319). While Brutus may commend his newfound vigour, this detail has little bearing on the plot other than to show how the co-conspirators perceive even issues as visceral as sickness or health as liable to be transformed by discourse.

While words can do much in this play, however, the most effective persuasions occur when visual and auditory cues meet, and specifically when bodily signs support rhetorical argument. Addressing the populace after Caesar's murder, Brutus believes that appeal to his own character will continue to work. He employs the same circular logic and language with which Cassius convinced him in 1.2 when he asserts that the assembled crowd will 'believe me / for mine honour, and have respect to mine honour, that / you may believe' (3.2.15–17). Though initially successful, he fails to account for the malleability of his audience, and is rhetorically outgunned by Anthony's use of the emperor's corpse as a character which can be made to narrate its own injuries:

[41] Xinyao Xiao, "Oxymoronic Ethos: The Rhetoric of Honor and Its Performance in Shakespeare's *Julius Caesar*". *Philological Quarterly* 97, no. 3 (2018): 263–85, (264).

[42] Miles Geoffrey, *Shakespeare and the Constant Romans* (Oxford University Press, 1996), 137.

> Through this [cloak] the well-beloved Brutus stabb'd;
> And as he pluck'd his cursed steel away,
> Mark how the blood of Caesar follow'd it,
> As rushing out of doors, to be resolved
> If Brutus so unkindly knock'd, or no. (3.2.176–80)

Prosopopoeia—that is, 'the faining of a person' in an inanimate object or an abstract concept—is employed here to masterful effect.[43] 'To stir men's blood', Anthony announces, he will 'Show you sweet Caesar's wounds, poor poor dumb mouths, / And bid them speak for me' (3.2.219–21). His suggestion that skilled oratory might 'put a tongue / In every wound of Caesar' is grotesque, but it speaks to the way in which rhetoric invests the matter of the body with meaning, and the body's materiality in turn gives rhetoric the ring of truth (3.2.223–24). Indeed, Gavin Alexander notes that 'Quintilian … actually used the events on which Shakespeare based his scene as an example when recommending that in the peroration at a murder trial the prosecutor might show fragments of blood-stained clothing or even the corpse itself'.[44] The encounter is given an extra *frisson* for Shakespeare's audience because of the Christological imagery attached to Caesar's body. The stab wounds therein become sources of nourishment as well as mortal injuries, and the blood that issues out makes potential relics: 'they would go and kiss dead Caesar's wounds / And dip their napkins in his sacred blood, / Yea, beg a hair of him for memory' (3.2.677–79). In Daniela Carpi's words, this body 'becomes the sacrificial object at the center of a profane mass' in which the onlookers 'eat' (touch) the corpse, which becomes a means of reconciliation between the ruler and the people.[45]

While Anthony's use of Caesar's body as a rhetorical tool has been widely discussed by critics, few have acknowledged that this tactic has

[43] Henry Peacham, *The Garden of Eloquence, Conteining the Most Excellent Ornaments, Exornations, Lights, Flowers, and Formes of Speech, Commonly Called the Figures of Rhetoricke.* (R.F. for H. Jackson, 1593), 136.

[44] Gavin Alexander, "Prosopopoeia: The Speaking Figure," in *Renaissance Figures of Speech*, ed. Katrin Ettenhuber, Sylvia Adamson, and Gavin Alexander (Cambridge University Press, 2007), 110.

[45] Daniela Carpi, "Caesar's Body in Shakespeare's *Julius Caesar*: Sacralization and De-Sacralization of Power". *Pólemos* 9, no. 2 (2015): 281–94. https://doi.org/10.1515/pol-2015-0019, 282.

already been rehearsed by Portia.[46] Far from merely foreshadowing Caesar's stabbing, Portia's self-wounding provides a pattern for the use of wounds as narrative tools. To look on Portia's wound, as she invites Brutus (and by extension the audience) to do, is to engage with the story to which it is attached. This storytelling takes place through the manual rhetoric of the wounding process, akin to gesture in its blend of spontaneity and referentiality. With reference to her wound, Portia is able to deploy well-established rhetorical tactics, including that of invoking one's lineage as support for claims about ethos: 'I grant I am a woman; but withal / A woman well-reputed, Cato's daughter'.[47] Cato was famed for having killed himself by tearing out his own entrails because he was unwilling to live under Caesar's autocracy. Though his suicide was troublingly violent and disorderly, it was, as Jenny Mayhew points out, 'unmistakeably rhetorical'.[48] Portia's status as Cato's daughter therefore helps to establish her 'situated ethos' (ethos gained through pre-existing social status). More specifically, it positions her self-wounding within a framework which sees the body as ideally subordinate to the mind, so that the infliction of pain is read as abnegation of the body in service of a higher goal rather than as mere hysteria.[49] Thus, though Portia's wound emblematises her emotional hurt, there is no suggestion that Portia stabs herself in order to relieve emotional distress or 'let out' her feelings, as is the case in some other representations of self-injury explored in this book. Rather, pain is an opportunity to demonstrate and cultivate that most prized attribute, constancy. In this respect, Portia echoes Cassius's show of daring in 'submitting' himself 'unto the perilous night' (2.3.46–52). Unlike Cassius, however, she follows through, and her action goes beyond risking harm to actually inflicting it.

* * *

[46] David Lee Miller, "The Voice of Caesar's Wounds: The Politics of Martyrdom in Shakespeare's Julius Caesar," in *Forms of Association: Making Publics in Early Modern Europe*, ed. Paul Yachnin and Marlene Eberhart (University of Massachusetts Press, 2015), 237; Carpi, "Caesar's Body".

[47] Xiao, "Oxymoronic Ethos", 266.

[48] Jenny Mayhew, "Godly Beds of Pain: Pain in English Protestant Manuals (ca. 1550–1650)," in *The Sense of Suffering: Constructions of Physical Pain in Early Modern Culture*, ed. Jan Frans van Dijkhuizen and Karl A. E. Enenkel (Brill, 2008), 307.

[49] Marshall, "Portia's Wound, Calphurnia's Dream", 474.

This begs a question: If the pain of wounding is an opportunity to demonstrate her constancy, why does Portia's stabbing take place off stage? In putting the act of voluntary wounding out of the audience's view, Shakespeare reduces its spectacle, though there is still considerable tension in the moment Portia reveals her injured thigh. However, it also affords Portia the opportunity to control the narrative which attends her wounding more closely than might otherwise be the case. As I demonstrate below and elsewhere in this book, representations of self-wounding often showed how the meanings intended by the protagonist could be lost in translation. There was an ever-present and worrying potential for wounding to be interpreted as pitiful, ignoble or simply foolish. Portia, however, is able to meticulously direct the revelation of her wound so that the injury appears at the height of her dispute with Brutus, and immediately after she has detailed all the ways in which *his* gestures tell the truth about his discontent: 'you suddenly arose, and walked about … stared … scratched your head … stamped with your foot' (2.1.238–43). She has already attempted to utilise her body to make her point by appealing 'upon my knees' to her 'once-commended beauty' (2.1.269–70), but uncovering her wound shows that she has anticipated how this conversation will go, and is prepared to out-argue her husband. Thus, the wound's political and ethical persuasiveness accrues from its combination of reference to a heroic model, formulaic display of constancy, and emotional context. 'Here in the thigh' is proof of genuine physical pain, of which the meaning is carefully stage-managed (2.1.300).

While the intended meaning of Portia's wound may be painstakingly elucidated, the text tells us tantalisingly little of its specifics. Is it a cut or stab? Of what depth? Is it still bleeding, or skinned over? Partly because of this ambiguity, critical readings of the voluntary wound have speculated widely on its sexual significance. The erotic charge of an intimately placed orifice which only one's husband may see is self-evident, as is the ironic potential of the way that opening the body finally enables it to be trusted as a closed container for secrets. Various critics thus interpret the wound as signifying both/either the vagina and/or a phallus, conferring masculinity or accentuating feminine vulnerability. Reading *Julius Caesar* through a Bakhtinian lens, Gail Kern Paster shows how the play's concern with bloodshed reflects contemporary anxieties about the feminine body as leaky, uncontained and volatile. By contrast, she argues, Portia's decision to voluntarily shed blood offers a means to humorally regulate the body: 'To have excess blood to shed … does not create gender difference;

what does is the possibility of shedding it *at will*. The male subject can regard such an action, therapeutically, as purgative and thus define it as enhancing rather than endangering somatic integrity'.[50] While Portia creates a wound that can be read as symbolising the vagina, she seeks to rewrite that signification by appealing to the reputation for honour which she has in common with Brutus:

> Portia calls attention to this bodily site not to remind Brutus of her femaleness, her lack of the phallus, but rather to offer the wound as a substitute phallus. Hers is not the involuntary wound of the leaking female body but the honorifically gendered, purgative, voluntary wound of the male.[51]

Paradoxically, a voluntary wound comes to stand for the impenetrability of the body, allowing Portia to move from the feminised domestic sphere into the masculine world of politics.

Paster's reading is convincing, but not final, for even when she commits manly acts, Portia remains something apart from her male counterparts. As Khan points out, though self-inflicted wounds serve as markers for masculine values, they inevitably also highlight the penetrability which is gendered feminine, 'making Portia's wound ambiguously, undecideably feminine and masculine'.[52] *Julius Caesar* is frequently concerned with the tensions created when apparently binary qualities co-exist: public and private; inside and outside; political and domestic; nature and culture. In this light, Portia's persuasive project is not a matter of refiguring her body as much as one of pluralising its potential meanings. Her body is still read according to formulas which designate it as having 'natural' predispositions, but those dispositions now include hereditary stoicism and valour alongside the supposed weakness of her sex. Moreover, the very act of using her body as a persuasive tool aligns her with male forms of power, specifically the power of rhetoric exercised by Cassius, Brutus and Anthony. As Kahn has highlighted, in establishing Brutus as the matter to be persuasively worked upon, Portia intrinsically casts herself as agential, virile and full of 'mettle'.[53]

[50] Gail Kern Paster, "'In the Spirit of Men There Is No Blood': Blood as Trope of Gender in *Julius Caesar*". *Shakespeare Quarterly* 40, no. 3 (1989): 284–98. https://www.jstor.org/stable/2870724, 288. My italics.

[51] Paster, "In the Spirit of Men", 294.

[52] Kahn, *Roman Shakespeare*, 101.

[53] Kahn, *Roman Shakespeare*, 77–109.

Shakespeare's interest in the potential of a voluntary wound as a persuasive tool is all the more visible when we consider the difference between this character on the Renaissance stage and in the source text, Plutarch's *Life of Brutus*. Shakespeare's Portia is certain of her own fortitude: 'Tell me your counsels; I will not disclose 'em' (2.1.297). Her self-injury is entirely outward-facing, undertaken with the explicit aim of convincing a taciturn husband. By contrast, Plutarch's 'Porcia' wounds herself in order to try her fortitude before approaching Brutus:

> Porcia ... did not try to question her husband about his secrets until she had put herself to the following test. She took a little knife, such as barbers use to cut the finger nails, and after banishing all her attendants from her chamber, made a deep gash in her thigh, so that there was a copious flow of blood, and after a little while violent pains and chills and fever followed from the wound. Seeing that Brutus was disturbed and greatly distressed, in the height of her anguish she spoke to him thus: "Brutus, I am Cato's daughter, and I was brought into thy house, not, like a mere concubine, to share thy bed and board merely, but to be a partner in thy joys, and a partner in thy troubles. ... Before this I put less confidence in these advantages, but now I know that I am superior even to pain."[54]

The process Porcia describes here of inuring herself to pain was, as Mayhew has described, familiar to early modern readers, and was sometimes recommended by Protestant handbooks.[55] Indeed, only a few years before the first performance of *Julius Caesar*, William Perkins's *Salve for a Sicke Man* described approvingly how the imprisoned Protestant Thomas Bilney had tested his ability to endure pain prior to his execution by repeatedly putting his finger in a candle flame.[56] Given the ubiquity of this idea, then, it is significant that Shakespeare chooses to entirely jettison the idea of self-stabbing as inurement in order to focus on the wound's communicative status.

For Shakespeare's Portia, self-wounding as a persuasive strategy seems to bear fruit. Her stated aim throughout the altercation has been to

[54] Plutarch, *Plutarch's Lives, Volume 2*. Revised from a 1683 translation by John Dryden (Random House Publishing Group, 2011), 580.

[55] Mayhew, "Godly Beds of Pain", 317–18.

[56] William Perkins, *A Salue for a Sicke Man: Or, A Treatise Containing the Nature, Differences, and Kinds of Death; as Also the Right Manner of Dying Well* (John Legatt, 1625), 83.

reinstate herself and her husband as a single psychological and physiological unit, 'incorporate and make us one /... your self, your half' (2.1.272–73). The bonds of marriage, she argues, stipulate that they should be one self, without 'limitations' (2.1.282). As such, Brutus' reaction to her display is to declare himself moved not only to admiration ('Render me worthy of this noble wife!' (2.1.302)) but to a new quasi-physical communion with his wife: 'by and by thy bosom shall partake / The secrets of my heart' (2.1.304–05). In a system where the physical and psychological components of emotion are indistinct, their reconciliation involves a mingling of their two substances, one's heart in another's bosom. Brutus promises that 'All my engagements I will construe to thee, /All the charactery of my sad brows' (2.1.306–07). In so doing, he draws on the plural meanings of 'construe' as both 'to expound' and 'to form by putting together materials'.[57] Henceforth, Portia's dramatic foregrounding of her individual body will give way to a union in which the 'charactery', or language, of their two bodies becomes as one.[58] Arguably, Portia's self-wounding also effects a broader emotional contagion which resonates throughout the play. Though other characters have threatened or foretold violence, Portia is the first to actually penetrate the bodily envelope, and in so doing she seems to open the flood gates. Following her self-stabbing, Caesar receives multiple stab wounds (3.1), Cassius narrowly avoids being stabbed by Brutus (4.2), Cassius persuades Pindarus to stab him (5.3), Titinius stabs himself (5.3) and finally Brutus runs on his own sword (5.5).

All this is not to say that the voluntary wound is unproblematic. The hard-won sense of shared existence which Portia has wrested from Brutus is fragile, and though it is implied that she will share Brutus's secrets, we never see this exchange take place. Before Brutus can explain himself to his wife, a knock at the door draws him back into the homosocial network within which his plot will be executed. More troubling is Brutus's reaction to news of his wife's death in 4.3. Brutus hears of the death at line 2145: 'No man bears sorrow better. Portia is dead'. By line 2160 he is requesting that Cassius 'Speak no more of her'. The stoicism which has framed a voluntary wound as proof of constancy in Portia is here exaggerated to the point of parody; a man happy to squabble with friends over petty matters but not to grieve the death of his spouse. At line 2196, however, he hears of the death again, from Messala. His response is again stoical: 'Why,

[57] "Construe, v." *Oxford English Dictionary*. https://www.oed.com/.

[58] "Charactery, n." *Oxford English Dictionary*. https://www.oed.com/.

farewell, Portia. We must die, Messala: / With meditating that she must die once, / I have the patience to endure it now' (2198–2200). Numerous critics have puzzled over Brutus's failure to reveal to Messala that he already knows Portia is dead. It is possible that Messala's announcement is a later textual revision and that the reviser (perhaps Shakespeare himself) simply failed to remove the reference at 4.3.2145. If we take the two incidents at face value, however, they indicate a troubling duplicity on Brutus's part; he is willing to lie by omission to bolster impressions of his own stoicism.[59] This strange turn of events therefore seems to have more to do with Brutus's personal shortcomings than it does with Portia's rhetorical prowess. By this point in the play, Brutus's capacity for self-delusion is abundantly evident, and he has too much invested in his persona of constancy to let it slip. For the audience, however, that investment actually undermines his claims to equanimity, since excessive attachment to honour, and therefore to the opinion of others, was identified by contemporary writers as antithetical to the ideals of stoicism.[60]

Notwithstanding the way in which it is reported, I contend that the manner of Portia's death vindicates rather than diminishes the force of her earlier self-wounding, by confirming her dedication to keeping her promises. Ever alive to the symbolism in her suffering, her dramatic suicide recollects that of her father Cato, following through on the inheritance which she highlighted as part of her persuasive ethos. Moreover, the image of ingesting hot coals once again draws on several cultural reference points. As I argue in the following chapter, self-injury of the tongue (albeit usually by autoglossotomy) was strongly associated in the early modern period with stoic ideals, particularly the ability to keep silent. Such a peculiar suicide also brings to mind the 'stomach' which was said (most famously by Elizabeth I) to underlie feats of courage, particularly masculine courage. Swallowing hot coals, Portia proves her ability to quite literally keep her mouth shut.

[59] Geoffrey, *Shakespeare and the Constant Romans*, 145.

[60] Geoff Baldwin explains how 'The valuing of honour and glory as an indication of worth had been questioned by scepticism ... In order to present a public persona given this atmosphere, it was important not to pour one's whole self into this persona, so that the individual could not be described by referring to the sum of duties or offices held ... Not only this, but the self which was reserved had to be capable of judgement, so that it could discern the deception of the world and direct the operation of one or more persona'. Baldwin, "Individual and Self in the Late Renaissance". *The Historical Journal* 44, no. 2 (2001): 341–64. https://www.jstor.org/stable/3133611, 364.

2.3 Self-Wounding as Object Lesson in *Tamburlaine the Great, Part II*

When Marlowe's eponymous hero (or anti-hero) Tamburlaine sets out to teach his sons a lesson, he does so against a backdrop of violence. In part one of his adventures, the low-born Tamburlaine is shown kidnapping a princess, imprisoning the Turkish emperor and conquering Africa. Part two sees him hell-bent on continuing to sack neighbouring kingdoms, unchecked by the death of his wife and hungry to recruit his sons as co-conquerors. First, however, he needs to teach them 'rudiments of war', from marching in armour to besieging a fort, digging ditches and strategically positioning one's soldiers (II, 3.2.54). The most important lesson they need to learn is never to be afraid, and for this Marlowe reserves the play's only example of violence against the mighty Tamburlaine:

> TAMBURLAINE: Villain, art thou the son of Tamburlaine
> And fear'st to die, or with a curtle-axe
> To hew thy flesh and make a gaping wound? /… /
> View me thy father that hath conquered kings
> And with his host marched round about the earth
> Quite void of scars and clear from any wound,
> That by the wars lost not a dram of blood,
> And see him lance his flesh to teach you all.
> *He cuts his arm*
> A wound is nothing be it ne'er so deep,
> Blood is the god of war's rich livery.
> Now look I like a soldier /… /
> Come, boys, and with your fingers search my wound
> And in my blood wash all your hands at once,
> While I sit smiling to behold the sight.
> Now, my boys, what think you of a wound? (II, 3.2.95–129)

Tamburlaine's self-wounding appears similar in many respects to Portia's, only bigger and bolder. Like Portia, he explicitly treats self-wounding as a tool for persuasion, meticulously stage-managing the context in which his wound is interpreted. True to form for a leader so invested in his capacity to shock and awe, the moment of his wounding is preceded by a twenty-line monologue, and he expounds the meaning of the lesson for a further fourteen lines before venturing to ask his sons 'what think you of a wound?' This overdetermination of the wound's meaning is compounded by a referential context which is both wider and more ambitious than Portia's evocation of her Stoic pedigree. Tamburlaine's principal use of his arm here is to evoke the military 'arms' on which his personal mythmaking project is based. He is constructed as an *übermensch* 'that hath conquered kings / And with his host marched round about the earth / Quite void of scars and clear from any wound' (II, 3.2.98–100). Wounding himself is an attempt (albeit ultimately unsuccessful) to reinforce this impression; shedding one's own blood maintains the myth that 'only Tamburlaine can wound Tamburlaine'.[61] This attempt to portray himself as simultaneously invulnerable and open also places Tamburlaine's cut, like Caesar's wounds, within the register of religious sacrifice. Numerous critics have noted that the emperor's invitation to his sons to 'with your fingers search my wound / And in my blood wash all your hands at once' recollects Jesus' appearance to St. Thomas after the resurrection.[62] Christopher Greenfield describes such resonances as 'blasphemous but not frivolous; they underline Tamburlaine's conviction of the sacredness of his person', while McEvoy and Hopkins usefully point out that Tamburlaine is already acquainted with the notion of self-wounding as piety, having in Part 1 described the 'priests of Mahomet' who cut themselves as a sacrifice (I, 4.2.2–3).[63]

Even as he invokes this sacrificial context, however, Tamburlaine takes the curious stance of attempting to prove his fortitude with reference to a wound that he insists he does not feel. Whereas Portia embraces her pain and uses it as proof of personal authenticity, Tamburlaine positively rejects bodily experience. Both the language of 'lancing' and the invitation to his sons to 'search' his wound evoke a surgical procedure, through which the patient sits uncannily 'smiling', as unfeeling as a corpse under the anatomist's knife. This disregard for the somatic is a stark contrast to Marlowe's other wounded characters, who, as Greenfield notes, 'speak vividly and compellingly about what they feel', including not only their pain but the processes occurring inside their body.[64] Cosroe, for example, narrates his

[61] Matthew R. Martin, *Tragedy and Trauma in the Plays of Christopher Marlowe* (Routledge, 2016), 74.

[62] Andrew Duxfield, *Christopher Marlowe and the Failure to Unify* (Ashgate Publishing, Ltd., 2015), 62–63.

[63] Matthew Greenfield, "Christopher Marlowe's Wound Knowledge", *PMLA* 119, no. 2 (March 2004): 233–246 (240); Lisa Hopkins and Sean McEvoy, *Christopher Marlowe, Renaissance Dramatist* (Edinburgh University Press, 2008), 124.

[64] Greenfield, "Christopher Marlowe's Wound Knowledge", 233.

own death (I, 2.7) for 18 lines, as 'what should be hidden in clouds of simple pain has the strongly etched lines of a Vesalian illustration'.[65] Tamburlaine, however, sees his cut in terms of that which he knows best—foreign treasure. His wound is, he insists,

> As great a grace and majesty to me,
> As if a chair of gold enamelled,
> Enchased with diamonds, sapphires, rubies,
> And fairest pearl of wealthy India,
> Were mounted here under a canopy,
> And I sat down clothed with the massy robe
> That late adorned the Afric potentate. (II, 3.2.118–24)

Tamburlaine here attempts the complete aestheticisation of his body and its suffering. The combination of somatic and symbolic which underpins manual rhetoric is strained to its utmost as he insists that he be interpreted in purely emblematic terms.

The outcome of this experiment in persuasion is mixed. While the two elder sons, Amyras and Celebinus, eagerly embrace their 'lesson', the youngest child, Calyphas, is less convinced: 'I know not what I should think of it. Methinks 'tis a pitiful sight' (II, 3.2.130). Unlike his brothers, Calyphas recognises that Tamburlaine's bombastic rhetoric cannot exist in a vacuum, but requires the participation of those it is designed to impress. While Calyphas recognises the intended significance of the wound, he also sees that he may reject that interpretation, insisting instead that the materiality of the body remains important and the sight of bodily harm is 'pitiful' however it may be framed. Articulating this uncertainty also aligns Calyphas with a theatre audience who, having seen the violent highlights of Tamburlaine's military career, may be similarly ambivalent. Tamburlaine's wound is couched in terms of his ability to control his body, but as Duxfield notes, his claim that 'Now look I like a soldier' has a 'demystifying' effect: 'Christ's wounds prove that he is the risen saviour, but Tamburlaine's emphasise his mortality, something which up to this point an audience may have doubted'.[66] Meg Pearson goes further in arguing that 'The spectacular wound Tamburlaine presents, when seen in the context of the

[65] Greenfield, "Christopher Marlowe's Wound Knowledge", 236.
[66] Duxfield, *Christopher Marlowe and the Failure to Unify*, 62–63.

ceremony he evokes, does not register as warlike. He wants to look like a soldier … but his ceremony casts him as the sacrificial victim instead'.[67]

This ambiguity is part of an alternative narrative in which Tamburlaine's self-wounding marks not self-mastery but an increasing loss of control. In Part I, the charismatic force of Tamburlaine's appearance and oratory is as powerful as his military genius. Meeting him as a 'Scythian shepherd' in 1.2, for instance, Theridamus comments that 'His fiery eyes are fixed upon the earth / As if he now devised some stratagem, / Or meant to pierce Avernus' darksome vaults/ And pull the triple-headed dog from hell' (I, 1.2.159–62). He is quickly won over by Tamburlaine's rhetoric: 'Not Hermes, prolocutor to the gods, / Could use persuasions more pathetical' (l. 210–11). Tamburlaine's short-lived alliance with Cosroe is made through similar means, as Menaphon describes in awed terms how each part of Tamburlaine's body 'Betokening valour and excess of strength / In every part proportioned like the man / Should make the world subdued to Tamburlaine' (I, 2.1.29–31). Moreover, Tamburlaine understands the effective use of visual cues as a part of persuasive strategy. When he lays siege to the Egyptian Soldan, his 'frowning looks' and 'terror and imperious eyes' strike fear into his opponents (I, 4.1.13–14). He even deploys a codified system of visual warnings to his enemies. At the beginning of a siege, we are told, Tamburlaine erects white tents 'To signify the mildness of his mind / That satiate with spoil refuseth blood' (I, 4.1.52–53). If this fails to work, the second day's tents will be red, signalling his intent to kill 'any that can manage arms' (I.4.1.57). Thereafter, he promises to kill men, women and children alike, and appears entirely clad in black, from his tents to 'His spear, his shield, his horse, his armour, plumes' (I, 4.1.60).

While Part II contains moments of similarly compelling spectacle, it is clear by the scene of the self-wounding that, as Pearson suggests, 'what was once a spectacular seduction is now a ham-fisted onslaught'.[68] The death of Zenocrate is a typical example of Tamburlaine's urge to pile rhetoric upon rhetoric, overdetermining events in order to make them work to bolster his waning authority. His response to bereavement includes, as Pearson lists, 'a memorial pillar written in Arabic, Hebrew and Greek forbidding the rebuilding of the burned town; a streamer bearing Persian and Egyptian crests to signify Zenocrate's royal birth; a tablet covered with

[67] Meg Pearson, "'Raving, Impatient, Desperate, and Mad': Tamburlaine's Spectacular Collapse". *Marlowe Studies* 2 (2012): 87–102 (94).

[68] Pearson, "Raving, Impatient and Mad", 87.

lists of her great qualities; the golden hearse itself; and finally, a picture of the dead queen'.[69] Meanwhile Zenocrate's only request, that Tamburlaine refrain from responding to her death with violence, goes ignored. Indeed, only a few scenes later he kills Calyphas, the son arguably most like his mother, in a display of violence which shocks even his allies and during which the repeated interjections of Theridamus, Techelles and Amyras show how far the leader has departed from inspiring obedience through sheer majesty. Viewed in this light, Tamburlaine's self-wounding forms part of a trajectory in which his ever-widening sphere of influence accompanies a diminution of the personal mythology at the centre of his empire.

Why does Tamburlaine's persuasion fail where Portia's succeeds? Clearly, there are multiple factors linked to his outsize personality—not least the appalling cruelty which renders sympathy with Tamburlaine almost impossible. At root, however, I argue that Tamburlaine's self-wounding fails to fully convince his audience because he neglects a foundational principle of rhetoric as described above; that one should actually feel the emotions one wishes to stir in others. As Alexander describes, rhetorical theorists fostered a notion of the ideal rhetor formulating his own ethos partly through the act of prosopopoeia; 'being able to move his audience *because* he has moved himself, and moving himself not so much by the force of actual recollection as by the power of the words he has written and is uttering'.[70] Tamburlaine, however, wishes only to inspire awe in his audience, while continuing to insist that he feels no pain. In so doing, he shows that he has only partly understood how self-wounding works as a rhetorical strategy. He has comprehended the attention-grabbing potential of a self-inflicted wound. However, he presents that wound as entirely symbolic in nature. If the damage inflicted on the body is as insignificant and painless as Tamburlaine insists, then self-injury functions *only* at the level of gesture. *Julius Caesar*'s Portia realises that pain is the 'strong proof' of her claims to constancy, providing somatic evidence of her authenticity. Tamburlaine fails to grasp this part of the equation, and thus remains vulnerable to the accusations of fakery which I have shown were levelled at rhetoricians and actors alike.

[69] Pearson, "Raving, Impatient and Mad", 90.

[70] Gavin Alexander, "Prosopopoeia: The Speaking Figure" in *Renaissance Figures of Speech*, ed. by Sylvia Adamson, Gavin Alexander, and Katrin Ettenhuber (Cambridge University Press, 2007), 101.

Tamburlaine's failure to grasp that the materiality of self-injury matters as much as its symbolism is not merely an oversight. Rather, it is consistent with the conqueror's general lack of a credible ethos to which he can appeal to prove his sincerity. What are Tamburlaine's 'real' feelings? Does he even have any? As Tromly has observed, all his interests tend towards acquiring what others have. Even his love of treasure is outward-facing, since the objects he attains have little value but that which is attached to them by others. In taking and bestowing territories and crowns, 'Tamburlaine contrives to turn people who are under his control into caricatures of himself; his manipulation of their hopes and desires creates a fun-house mirror in which his own failure to find satisfaction is reflected'.[71] With the possible exception of Zenocrate, who seems as much a favourite possession as a wife, he also fails to form the intersubjective bonds so important to early modern conceptions of identity. Keeping even his closest allies as subordinates, he accepts his sons as his own only if they parrot his ambitions: 'Be all a scourge and terror to the world, / Or else you are not sons of Tamburlaine' (II, 1.4.63–64).

Whether Tamburlaine's self-injury proved more impressive to theatrical audiences was unclear. As Tom Rutter points out, Marlowe was unusual in inviting audiences of the play to judge it 'as you please', and determinedly avoiding a clear moral lesson in either part of the drama. Playgoers responded enthusiastically; Dawson finds that '*Tamburlaine* was one of the most popular plays of those years [1580s and 1590s], a fact that can be gauged not just from frequency of performance but from the gate receipts also recorded by Henslowe'.[72] Nonetheless, Marlowe's propensity for 'waving his rhetorical talent like a banner' also attracted criticism, primarily from other writers.[73] In the epistle prefacing *Perimedes the Blacksmith* (1588), Robert Greene complained that *Tamburlaine*'s verse, with 'everie word filling the mouth like the faburden of Bo-Bell', made 'intolerable poetrie'.[74] Ben Jonson likewise selected the play for particular censure

[71] Fred B. Tromly, *Playing with Desire: Christopher Marlowe and the Art of Tantalization* (University of Toronto Press, 1998), 68.

[72] Anthony Dawson, 'Introduction' in Marlowe, Christopher *Tamburlaine The Great: Part Two* (Bloomsbury Publishing, 2003), xxix.

[73] Russ McDonald, "Rhetoric and Theater" in *The Oxford Handbook of Rhetorical Studies* (Oxford University Press, 2017), 454.

[74] Robert Greene, *Perimedes the Blacke-Smith a Golden Methode, How to use the Minde in Pleasant and Profitable Exercise* (John Wolfe, for Edward White, 1583), fol. A3r.

when he complained about overblown rhetoric in his 'Timber, or Discoveries':

> The true Artificer will not run away from nature, as hee were afraid of her; or depart from life, and the likenesses of Truth; but speak to the capacity of his hearers, and though his language differ from the vulgar somewhat; it shall not fly from all humanity, with the Tamerlanes, and Tamer-Chams of the late Age, which had nothing in them but the scenicall strutting, and furious vociferation, to warrant them to the ignorant gapers.[75]

Jonson's invective was directed equally at the 'ignorant gapers' who craved bombast and at Marlowe, who supplied it. Such critiques were levelled at many of Marlowe's works; Catherine Nicholson records how

> Thomas Nashe heard in Marlowe's verse both ill-disguised insufficiency—"the swelling bombast of a bragging blanke verse"—and blatant excess—the "ingrafted overflow" and "spacious volubility of a draumming decasillabon"; while Joseph Hall dismissed Marlowe's "pure Iambick verse" as a far-fetched concoction "patch[ed] … up" with "termes Italianate".[76]

In the case of *Tamburlaine*, however, Jonson and others may have (perhaps purposely) missed the point. Marlowe makes no claims to subtlety, but rather shows Tamburlaine's rhetoric to be overblown, not least through the spectacle and ultimate failure of his self-wounding. In pushing the connection between bodily experience and symbolic meaning to the point of failure, Marlowe invites audiences to consider how meaning is made with, and through, the body.

2.4 Conclusion

As a piece of communicative practice, early modern self-injury was uniquely positioned to convince audiences of the practitioner's integrity, their status as *just what they claimed to be*. By literalising the violence of a person's feelings or desires, self-wounding manifested a foundational principle of

[75] Ben Jonson, "Timber, or Discoveries Made upon Men and Matter," in *The Works of Ben. Jonson.*, vol. 7 (printed for D. Midwinter; W. Innys and J. Richardson; J. Knapton; T. Wotton; C. Hitch and L. Hawes, 1756), 95.

[76] Catherine Nicholson, *Uncommon Tongues: Eloquence and Eccentricity in the English Renaissance* (University of Pennsylvania Press, 2013), 143–44.

rhetoric; that 'personating' a feeling, whether through words or movements, involved really experiencing that feeling. If one used self-stabbing to convince one's husband of fortitude in the face of hardship, the persuasive act itself involved fortitude.

This did not mean that wounding oneself was an automatic route to authenticity. As the differences between Portia and Tamburlaine's on-stage self-injuries show, this, like other kinds of rhetoric, was constrained by environmental and social context. If it was not made readable by careful contextualisation, self-injury risked being interpreted in unintended ways. As the next chapter will demonstrate, this danger was highlighted in Thomas Kyd's *The Spanish Tragedy*, where small textual changes could change the meaning of the protagonist's self-injury from heroic to disturbing. Conversely, focussing too much on the self-injury as a symbolic spectacle, and denying its materiality, also risked undermining the practitioner's claims to authenticity. If it was not painful, then a self-inflicted wound had no more claim to veracity than any other kind of gesture. It was as hollow as some twentieth-century psychologists found the 'suicide gesture', an act lacking 'real' intent.

Renaissance playwrights thus used self-injury as one means to worry at the question of how one could ever establish another person's authenticity, or prove one's own authenticity to others. Though these issues remained unresolved, they seemingly struck a chord with audiences, and later chapters of this book will show that even off-stage, self-wounding was positioned as theatrical. Scenes of 'discovery' of the wounded person often seemed tableau-like, while the narratives which explained such events were neatly constrained and sometimes blackly humorous. The principles established by theories of gesture and rhetoric found new expression in the spectacle of on-stage self-wounding, and they would go on to permeate discussion of self-injury in every other kind of early modern text.

CHAPTER 3

Protest: Autoglossotomy and Biopolitics in *The Spanish Tragedy*

OLLIE: I'd like us to start with a question… Why do we think people self-injure?
[Everyone pauses and looks around uncertainly]
TAMBURLAINE: For me, it made me feel more powerful.
CARLOS: Powerful? It doesn't make me feel powerful. Sometimes, the only way I can distract myself from my mental pain is to create physical pain on myself. I can understand it more if I can see it this way.
BEN: I know what you're saying, I've never seen it as powerful before, I see it as vulnerability.
TAMBURLAINE: For me, there was an element of performance. In my time, we had to make a statement to be listened to.[1]

Thus far, I have shown how self-wounding might function as a communicative practice, utilising a combination of formulaic meanings and perceived naturalness to persuade audiences of a practitioner's authenticity. This chapter develops that theme by considering further what meanings were attached to self-wounding, and what happened when these meanings met and clashed. In Thomas Kyd's *The Spanish Tragedy*, self-injury may be read as heroic, desperate or insane. Through the morally ambiguous figure of Hieronimo, extreme self-inflicted violence becomes a means to consider what is now termed 'biopower', and in turn, to highlight anxieties about contemporary systems of justice and the ethics of staging bodily harm as theatrical spectacle.

[1] Quote from *I Am Worthy,* play devised by participants in the Self Injury Through Time theatre group, 2024.

A. Skuse, *Hurt Feelings*, Palgrave Studies in Literature, Science and Medicine, https://doi.org/10.1007/978-3-032-04565-2_3

The first part of this chapter argues that while early modern people did not find a causal relationship between self-injury and suicide (as is usually the case in modern discourses on NSSI), self-wounding was nonetheless conceptually linked to self-killing. Whilst suicide and self-wounding were legally and culturally distinct, both acts could be viewed as an assertion of ownership over the body, whether for good or ill. In particular, neo-stoic writings construct pain as an opportunity for personal growth and therefore present self-wounding as a potential means of self-fashioning. The latter parts of this chapter show how these debates might play out on the early modern stage by examining Thomas Kyd's *The Spanish Tragedy*, a play in which self-wounding is overburdened with potential meanings. In this prototypical revenge tragedy, the protagonist Hieronimo bites out his own tongue purportedly to avoid giving up his secrets, but the irony is that he has already revealed his plot—in fact, he revels in its revelation. I argue that this apparent puzzle is solvable if we consider Hieronimo's autoglossotomy as an act of resistance which utilises ideologies attached to self-wounding in general and beliefs about tongues and tongue-biting in particular. Viewed in this way, Hieronimo's self-wounding is shown to function as an equal and opposite response to the threats of torture which threaten his bodily autonomy (and struck close to home for the early modern audience). Indeed, it bears striking similarities to ideas of biopower or 'necroresistance' as they are presented in modern cultural studies. However, later versions of the play open up other interpretive possibilities for his autoglossotomy. By suggesting that Hieronimo's actions are driven by insanity, these versions emphasise self-wounding as a disturbing spectacle which may as easily be read in terms of anguish as of agency. This may still be madness, even if there is method in it.

3.1 The Ethical Status of Self-Wounding

Since the twentieth century, self-injury has been deemed a significant risk factor for death by suicide. Some studies indicate that people who self-injure have an up to 40 per cent risk of attempting suicide, while others highlight the 'versatility' of self-injury activities (the number of methods used) as more accurately predicting the likelihood of suicide.[2] This clear

[2] Rebecca C. Groschwitz, Michael Kaess, Gloria Fischer, et al. "The Association of Non-Suicidal Self-Injury and Suicidal Behavior According to DSM-5 in Adolescent Psychiatric Inpatients". *Psychiatry Research* 228, no. 3 (2015): 454–61. https://doi.org/10.1016/j.

association—self-injury to self-destruction—may seem intuitive, but it is in stark contrast to early modern thinking on the topic. In many of the multiple interpretations of self-wounding in early modern texts, it seems to have surprisingly little to do with suicide. For example, medical and ballad accounts of men's self-castration make clear that those men were not viewed as having a death wish, however dangerous their endeavour. Likewise, women who tore out their hair did so as part of a lament tradition that helped ensure that after a period of mourning, life could carry on. Furthermore, self-wounding and suicide were legally distinct. Suicide was prohibited, and as I discuss below, it became subject to stricter, more rigorously enforced legal penalties in the sixteenth century. By contrast, self-wounding seems to have incurred no legal penalty.

The distinction between attempted suicide and non-suicidal self-injury apparently rested on the way in which the event was framed (by both protagonist and onlookers) rather than the seriousness of the harm. For example, Hannah Allen, who wrote of having been afflicted by deep religious melancholy in her 1683 *God's Gracious Dealings with the Choice Christian Mrs Hannah Allen*, recalled how in the depths of her illness she sought numerous ways to die. These included: collecting large spiders and smoking them in a pipe; attempting to procure an overdose of opium; taking a coach to Barnet Woods, planning to stay there until she died of exposure; cutting herself with scissors; and hiding under floorboards in her own home, planning to starve to death.[3] It was clear from Allen's narrative that many of these incidents caused fairly mild physical harm. She had not come close to death at any time. Yet, she made clear that these were still suicide attempts rather than incidences of self-injury. Conversely, the people I describe later in this book who injured themselves very

psychres.2015.06.019; Kerry L. Knox and Janis Whitlock. "The Relationship Between Self-Injurious Behavior and Suicide in a Young Adult Population". *Archives of Pediatrics & Adolescent Medicine* 161, no. 7 (2007): 634–40. https://doi.org/10.1001/archpedi.161.7.634; Brianna J. Turner, Sean M. Butler, Alexander L. Chapman, and Brianne K. Layden. "How Often, or How Many Ways: Clarifying the Relationship Between Non-Suicidal Self-Injury and Suicidality". *Archives of Suicide Research* 17, no. 4 (2013): 397–415. https://doi.org/10.1080/13811118.2013.802660.

[3] Hannah Allen, *A Narrative of God's Gracious Dealings with that Choice Christian Mrs. Hannah Allen (afterwards married to Mr. Hatt.). Reciting the Great Advantages the Devil Made of Her Deep Melancholy, and the Triumphant Victories, Rich and Sovereign Graces, God Gave Her over All His Stratagems and Devices.* (John Wallis, 1683).

seriously by cutting off their testicles were not thought to have acted from a desire to die, or to be at risk of completing suicide.

Self-injury and self-killing were therefore not *causally* linked in the way they commonly are today. The former was not thought to lead to the latter. However, this does not mean that they were completely divorced. Rather, they were *conceptually* linked, in that some of the moral arguments around suicide could also be applied to self-wounding. Evidence for this overlap comes principally from the seventeenth century, when the ethics of self-killing became a subject of intense interest.[4] In the glut of religious and moralistic writings on suicide printed during this period, self-wounding lingers at the margins of the discussion. A 1647 text, for instance, warned that 'It is dangerous to deface the work of God', as well as 'to unmake that which God hath made'.[5] Another, published in 1652, argued that the commandment against killing forbade men not only to actually 'take away the life' but also 'to hinder the health, and strength, and to hasten the death of themselves, or other men'.[6] It became a common theme among moralising authors to characterise self-neglect through unhealthy lifestyles as slow means to suicide, arguing that 'God may justly challenge them of self murder, that thus will eat and drink in spite of health and nature, and so shorten that precious time, which God indulged for better purpose'.[7] In a mid-seventeenth-century table showing varieties of self-murder, William Slatyer included neglect of one's physical or spiritual welfare, deliberate sinning, and 'Thrusting one's self in danger' alongside the more obvious means of 'committing some capital crime' or simply 'Being their own butchers, & murderers'.[8]

Such arguments stretched the notion of self-destruction to breaking point as a rhetorical tactic. But an explicit and serious link between

[4] See for instance Georges Minois, *History of Suicide: Voluntary Death in Western Culture* (John Hopkins University Press, 1999), 149; Marzio Barbagli, *Farewell to the World: A History of Suicide* (John Wiley & Sons, 2015), 82.

[5] Joseph Caryl, *An Exposition with Practicall Observations Continued upon the Eighth, Ninth and Tenth Chapters of the Book of Job* (A. Miller, for Henry Overton and Luke Fawne and John Rothwell and Giles Calvert, 1647), 496.

[6] L. S, *Natures Dowrie: Or The Peoples Native Liberty Asserted* (W.R., 1652), 21.

[7] John Toy, *A Sermon Preached in the Cathedrall Church of Worcester the Second of Febr. Last Being Candlemas Day, at the Funerall of Mris Alice Tomkins Wife unto Mr Thomas Tomkins One of the Gentlemen of His Majesties Chappell Royall* (S.N., 1642), 7.

[8] William Slatyer, *The Compleat Christian, and Compleat Armour and Armoury of a Christian, Fitting Him with All Necessary Furniture for That His Holy Profession, or, The Doctrine of Salvation* (London, 1643), 385.

self-wounding and suicide was articulated most clearly in the first published text devoted to the subject, John Sym's 1637 *Life's Preservative Against Self-Killing*. Sym posited that

> a man may be guilty of self-murder, by needlesse mutilating of himselfe, and cutting off any of his members, (as Origen did) to the hurt and danger of his life; which, by the preservation of such a member, might have beene in more safety; for, life's perfection is in the perfection of the whole body. Notwithstanding, for the safety of the whole, a man may lawfully and necessarily cut off a member; which cannot be preserved without manifest danger of thereby losing his life; but, neither to punish a sin past, nor to prevent a sin to come, may a man destroy or cut off any of his members, whereby he may be lesse able to doe the offices and duties for which God hath given him the same.[9]

Notably, Sym used as an example Origen of Alexandria, who castrated himself to avoid accusations of impropriety with female students, but clearly *without* suicidal intent.[10] Origen's actions were, he argued, sinful both because they damaged the body which was in the image of God, and because they risked death even if that was not their intended purpose. In more strident language, Alexander Seton's 1699 text on mutilation similarly warned that '*Castratio viridium* is one of the most atrocious *Demembrations*; and when a man does it to himself, he is *sui homicida* … And so punishable with Death and Confiscation of Goods … And its [sic] equivalent if one suffered himself willingly to be castrated by another'.[11]

This view was shared, albeit in more nuanced terms, by John Donne, whose *Biathanatos* was written in 1608 but considered so controversial

[9] John Sym, *Lifes Preservative against Self-Killing. Or, An Useful Treatise Concerning Life and Self-Murder* (M. Flesher, 1637), 110.

[10] See John Anthony McGuckin, *The Westminster Handbook to Origen* (Westminster John Knox Press, 2004); Bonnie Wheeler, "Origenary Fantasies: Abelard's Castration and Confession". In *Becoming Male in the Middle Ages*, ed. Jeffrey Jerome Cohen and Bonnie Wheeler (Garland Publishing, 1997).

[11] Alexander Seton, *A Treatise of Mutilation and Demembration* (Andrew Symson, 1699), 16.

Interestingly, this sentiment also appeared in some medical texts, as when the midwife Jane Sharp noted that it was unacceptable to try and make oneself infertile: 'for men or women to mutilate themselves on purpose, or use obstructive means to cause barreness … I can not think to be justifiable, though some persons have presumptuously ventured upon it'. Jane Sharp, *The Midwives Book* (Simon Miller, 1671), 164–65.

that it remained unpublished until 1647/8.[12] Donne's beliefs about suicide were far from clear-cut, but he was unambiguous in making a connection between self-wounding and self-killing:

> though in Civill courts, it be not subject to like penaltie, yet if it bee accompanied with the same malignitie, *it is in conscience the same sinne*, especially towards our selves, because it violates the same reason, which is that none may usurpe upon the bodie over which he hath no Dominion.[13]

While 'the same malignitie' might be interpreted as meaning suicidal intent, the examples Donne uses to illustrate this point suggest otherwise. For instance, he noted that the fourth-century Apostolic Canons expressly forbade men who had castrated themselves from being inaugurated as priests, stating that 'He that gelds himself cannot be a Clerk, because he is an [sic] Homicide of himself, and an enemy to God's creature'.[14] Moreover, Donne pointed out, Calvin implicitly condemned self-injury when he used it as a metaphor for the sinfulness of divorce: 'Calvine, esteemed [self-injury] so hainous, that he builds his Argument against divorce on this Ground [God made them one Body, and *it is in no case lawfull, for a man to teare his own body*]'.[15]

Texts such as Donne's and Sym's articulated the belief that self-wounding and suicide might be the same *sort of thing* because they both treated the body as belonging to the individual, in defiance of any higher authority. Condemnations of suicide rested largely, though not exclusively, on the argument that suicide usurped the right of life and death over the body which properly belonged to God, and through God's ordinance to the divinely appointed King or Queen. In 1673, for instance, J. Shafte wrote that 'the Right of Nature [over one's own life] is only a liberty that each man hath to use his power for his own preservation, not his own

[12] This caution was apparently justified: Philipot's *Self-homicide-murther* asserts that a rash of recent suicides may be attributable to reading of *Biathanatos* (Thomas Philipot, *Self-Homicide-Murther, or, Some Antidotes and Arguments Gleaned out of the Treasuries of Our Modern Casuists and Divines against That Horrid and Reigning Sin of Self-Murther* (W. Downing for Edward Thomas, 1674) fol. A2r).

[13] John Donne, *Biathanatos a Declaration of That Paradoxe or Thesis, That Selfe-Homicide Is Not so Naturally Sinne, That It May Never Be Otherwise: Wherein the Nature and the Extent of All Those Lawes, Which Seeme to Be Violated by This Act, Are Diligently Surveyed* (John Dawson, 1644), 133. My emphasis.

[14] Donne, *Biathanatos,* 133.

[15] Donne, *Biathanatos,* 134. Donne's parentheses, my emphasis.

destruction'.[16] John Adams's *Essay Concerning Self-Murther* similarly asserted that '*Life continued* is as fully God's Propriety, as *Life first bestow'd*; and therefore it must be an Act of the greatest Injustice for any Man to cut it off by Self-Murther'.[17] This reasoning echoes the much later notion of 'biopower', which I discuss below, and it underpinned not only the immorality but the illegality of suicide. Opponents of suicide pointed out that 'If Self-killing were lawful, the power of life and death immediately invested by God in the supream Magistrate [i.e. the monarch], would by degrees shrink to a despicable nothing'.[18] It is for this reason, argues Eric Langley, that 'The self-slaughterer is, we are repeatedly told, a dangerously antisocial individual'.[19]

As I shall discuss, early modern writers were well aware that suicide and self-injury could happen as consequences of melancholy and frenzy, and they urged sympathy for such cases. Robert Burton's *Anatomy of Melancholy*, for instance, advised that 'hard censures of such as offer violence to themselues, are to be mitigated, as in such as are mad, beside themselues, or knowne to have bin long melancholy, and that in extremity, they know not what they doe'.[20] However, in the late sixteenth and early seventeenth centuries, the threshold for judging a suicide to have been caused by madness was set almost impossibly high. In the medieval period, most suicides were deemed *non compos mentis*, with that verdict being handed down by local courts often composed of the victim's neighbours, and thus sympathetic to the plight of the family left behind. But Langley finds that between 1485 and 1660, government crackdown, and small rewards for coroners recording a *felo de se* verdict, meant that only two per cent of suicides were deemed *non compos mentis*.[21] Michael MacDonald and Terence Murphy likewise conclude that 'The intensification of hostility to self-killing in the sixteenth and early seventeenth centuries was the

[16] J. Shafte, *The Great Law of Nature, or, Self-Preservation Examined, Asserted and Vindicated from Mr. Hobbes His Abuses* (Printed by the author, 1673), fols. 2v–3r.

[17] John Adams, *An Essay Concerning Self-Murther* (Tho(mas) Bennet, 1700), 7; See also Ezra Pierce, *A Discourse of Self-Murder* (Rich. Baldwin, 1692), 15.

[18] Philipot, *Self-Homicide-Murther*, 9.

[19] Langley, *Narcissism and Suicide in Shakespeare and His Contemporaries*, 205.

[20] Burton, *The Anatomy of Melancholy*, 276–77.

[21] Eric Francis Langley, *Narcissism and Suicide in Shakespeare and His Contemporaries* (Oxford University Press, 2009), 212–13.

result of the Tudor revolutions in government and religion—the rise of the modern state and the Protestant reformation'.[22]

Civic and philosophical condemnations of self-killing rested on the view that suicide was an unethical usurpation of power over the body by an individual. It follows that in those instances where self-wounding was discursively linked with self-killing, it must have been deemed immoral on the same grounds. Therefore, even where the two kinds of activity were not explicitly connected, the emotions of fear and shame associated with self-killing may still have 'stuck' to self-wounding.[23] As I will show later in this book, shame certainly was a feature of certain kinds of self-wounding, specifically self-gelding. In one eighteenth-century report, for instance, a man who cut off his own testicles was afterwards so 'ashamed of his late enterprise, nor daring to stir out of the house, [he] removed his quarters by nights, and was never after heard of'.[24] However in other forms—particularly the 'heroic' sorts of self-wounding represented in dramatic texts—self-injury might even be a source of pride. Though this might appear contradictory, it reflects several factors. Firstly, self-injury encompassed a varied set of phenomena, and not all kinds of necessarily attracted the same associations with suicide. The highly stylised self-injury associated with lament, for instance, was seldom linked to discourses of self-killing. This was partly because violent lament practices such as hair-tearing caused comparatively mild physical damage, and partly because those practices emphasised self-injury as a means of emotional regulation aimed at carrying on with life, rather than the opposite. Conversely, self-gelding was, as we have seen, readily compared with suicide (I discuss some reasons for this in Chap. 5). Secondly, the discourses surrounding self-killing were themselves considerably less homogenous than they might appear. It is certainly true that religious writing reliably designated suicide as sinful and therefore shameful, and this designation dominated moralistic and civic discussions of the same. Nonetheless, as Drew Daniel has recently observed, imaginative representations of voluntary death might embrace the pleasure and even comedy to be found in this form of violence:

[22] Michael MacDonald and Terence R. Murphy, *Sleepless Souls: Suicide in Early Modern England* (Clarendon Press, 1993), 5.

[23] On emotions as potentially 'sticking' to certain objects or activities, see Sara Ahmed, *Cultural Politics of Emotion* (Edinburgh University Press, 2014).

[24] Daniel Turner, *The Art of Surgery: In Which Is Laid down Such a General Idea of the Same, as Is Confirmed by Practice*, 6th ed., vol. 2 (C. Rivington, and J. Clarke, 1741), 278–79.

> The intellectual history of early modern attitudes toward voluntary death has largely avoided a reckoning with what this book will show is in fact a pronounced contrary aesthetic tendency: *surges of positive affect, humor, and pleasure found within textual representations of self-harm, self-sacrifice, and violent death.*[25]

One of the principal bases for 'positive affects' around voluntary death in early modern literature, and particularly drama, was the philosophy of neo-Stoicism. Originally a philosophical school of ancient Greece, Stoicism gained a renewed following in Renaissance England with the translation in 1567 of Epictetus's *Enchiridion*. The central tenet of this work—that one should avoid worrying about things which cannot be controlled—proved so appealing to its audience that it helped establish a new genre in sixteenth-century publishing. A.A. Long asserts that 'No secular books were more widely read in the Renaissance than Cicero's *On Duties* (*De Officiis*), the *Letters* and *Dialogues* of Seneca, and the *Manual* of Epictetus'.[26] A flurry of translated and original works bolstered the movement, including Guillaume Du Vair's *A Buckler Against Adversitie* (1622), Tacitus's *The Annales* (1598) and Michel De Montaigne's essay 'In defence of Seneca and Plutarch' (1580). Perhaps most influential was Sir John Stradling's translation of Justus Lipsius's *De Constantia* in 1594. Marrying ideas from Tacitus and Seneca with Christian values, *De Constantia* was foundational to neo-Stoicism, a system of ethics which emphasised justice and constancy while leaving 'a considerable amount of room for the practice of pagan and Machiavellian statecraft'.[27] Moreover, Stoicism was particularly in vogue among writers. Daniel Cadman, for instance, finds that 'Neo-stoicism ... gained a following from a niche group of courtiers in the late Elizabethan and early Jacobean eras. Its popularity amongst members of the Sidney and Essex circles ensures it would have been familiar to the writers of the neo-Senecan dramas'.[28]

[25] Drew Daniel, *Joy of the Worm: Suicide and Pleasure in Early Modern English Literature* (University of Chicago Press, 2022), 10. [author's italics].

[26] A.A. Long, "Stoicism in the Philosophical Tradition: Spinoza, Lipsius, Butler" in *The Cambridge Companion to the Stoics*, ed. Brad Inwood (Cambridge University Press, 2003), 365.

[27] Theodore G. Corbett, "The Cult of Lipsius: A Leading Source of Early Modern Spanish Statecraft". *Journal of the History of Ideas* 36, no. 1 (1975): 139–52. https://www.jstor.org/stable/2709016, 143.

[28] Daniel John Cadman, "Republicanism and Stoicism in Renaissance Neo-Senecan Drama" (Ph.D., England, Sheffield Hallam University (United Kingdom), 2011), 18–19.

Part of the imaginative usefulness of Stoicism in this context lay in its association with political resistance, an association which would be put to work in *The Spanish Tragedy*. While Lipsius's work carefully advocated obedience to state authority, the Stoic's emphasis on pursuing virtue above all else naturally raised questions about where their allegiance lay. As Christopher Brooke explains,

> the perennial anxiety regarding those who present themselves as Stoics is that however assiduously they may be seen to be performing their various offices, they threaten to disrupt the smooth workings of a hierarchical political regime, substituting their own scale of values—regarding justice or honour, for example—for those that are officially recognised and socially sanctioned, and always placing their paramount concern with their own virtue above what others take to be the needs of the commonwealth for orderly, stable government.[29]

This was particularly troubling given the other feature of Stoicism: indifference to pain. Stoics did not deny the reality of pain sensations. Nonetheless, they counselled that pain was not bad in and of itself. Only virtue was truly good, and only vice was truly bad. All other circumstances which we might call good or ill—health, wealth, power—were either 'preferred indifferents' or 'unpreferred indifferents', which only acquired their moral character from the emotional judgements made about them. Feelings of pain, while real, were exacerbated by the emotional belief in pain's undesirability, and this response could be controlled. 'It is incredible', argued Du Vair, 'what power Reason and Discourse have … not onely to make us constant, but even to make paine appeare to us sweet and pleasant'.[30]

Such a view emptied pain of meaning, because pain was a 'false evil', a sensation to which opinions did not need to be ascribed.[31] However, it also empowered the individual to use pain for self-development: 'The Stoic individual, then, constitutes himself in part through pain; pain offers

[29] Christopher Brooke, *Philosophic Pride: Stoicism and Political Thought from Lipsius to Rousseau* (Princeton University Press, 2012), 74.

[30] Guillaume Du Vair, *A Buckler against Adversitie, or, A Treatise of Constancie* (Bernard Alsop, 1622), 32.

[31] Justus Lipsius, *Two Bookes of Constancie. Containing, Principallie, A Comfortable Conference, in Common Calamities. And Will Serue for a Singular Consolation to All That Are Priuately Distressed, of Afflicted, Either in Body or Mind*, trans. John Stradling (Printed by P. Short for Richard Johnes, 1595), 15.

an opportunity to exercise true self-control'.[32] Moreover, since bodily wellbeing was a preferred indifferent rather than a virtue in itself, it followed that if it were necessary to the preservation of their moral integrity, the Stoic could calmly and dispassionately bring about their own death:

> For that body is all that is vulnerable about me; with this dwelling so liable to injury there lives a spirit that is free ... I shall dissolve our partnership when this seems the proper course, and even now while we are bound one to the other the partnership will not be on equal terms: the soul will assume undivided authority. Refusal to be influenced by one's body assures one's freedom.[33]

In theory at least, this approach would mean that self-wounding was ethically permissible and even laudable if done in service of a higher political or personal goal. As one neo-Stoic text recommended, Stoics could 'use [the body] as a borrowed garment, to make shew for a time of our minde'.[34] In fact, as Daniel observes, it might be considered weak *not* to end one's life rather than submit to actual or figurative slavery; the 'self-possession' implied by self-killing was here not a usurpation of divine rights but the natural antithesis to enslavement.[35] As I discuss below, this notion of self-possession was strongly evident in stories about autoglossotomy. It also chimed with larger discourses about the ideal relationship between reason and passion. The notion of subordinating emotions or physical sensations to reason was by no means exclusive to Stoicism. Christopher Tilmouth has argued that throughout the Tudor period, schools of thought from Neoplatonism to Aristotelianism championed reason as that which 'intuits transcendental values, grasping a divinely sanctioned, eternally present morality which is the bond of all society'.[36] Nonetheless, this rationalist, neo-Stoic interpretive framework was one readily available to early modern audiences, who flocked to plays in which the heroes professed some version of Stoic ethics. Early modern writers

[32] Dijkhuizen and Enenkel, "Introduction", 15.

[33] Seneca, quoted in Sheetal Lodhia "'The House is Hers, the Soul is But a Tenant': Material Self-Fashioning and Revenge Tragedy". *Early Theatre* 12, no. 2 (2009): 135–61 (137).

[34] Du Vair, *A Buckler Against Adversitie*, 33.

[35] Daniel, *Joy of the Worm*, 14.

[36] Christopher Tilmouth, *Passion's Triumph over Reason: A History of the Moral Imagination from Spenser to Rochester* (Oxford University Press, 2011).

representing self-wounding thus found themselves with a surfeit of possibilities when it came to presenting the ethical dimensions of this phenomenon. Was self-wounding a heroic disavowal of the body's strictures, or a sinful insult to the authority of God and King? Was it perhaps an act to be pitied, just as some suicides were to be pitied for their insanity? In the hands of a skilled author, all these possibilities could be kept in play, with remarkable results.

3.2 Autoglossotomy as Agency in *The Spanish Tragedy*

In dense tomes such as Donne's *Biathanatos*, the ethics of self-wounding could seem as far removed from real life situations as treatises of rhetoric were from most people's everyday conversations. Once again, however, imaginative literature worked to turn these theoretical conundrums into dilemmas which were, if not relatable, then at least compelling. This was nowhere truer than in Thomas Kyd's immensely popular *The Spanish Tragedy*. Beloved by audiences for its twisting plot and self-consciously theatrical violence, the play was a box office hit, performed twenty-nine times by 1597 and published in ten quarto editions before 1633. Its reach was such that it is credited with influencing Marlowe, Tourneur, Middleton, Webster and Ford, and providing inspiration for Shakespeare's *Hamlet*.[37] It also offered an exploration of the way in which self-wounding might work as an assertion of Stoic self-ownership, as the protagonist Hieronimo commits a spectacular act of self-mutilation in direct response to the threat of torture.

Hieronimo's tongue-biting comes at a moment of crisis in a play built upon conflict. Against a backdrop of war between Spain and Portugal, the Spanish officer Don Andrea has been killed by the now-captive Portuguese Balthazar. Unable to be allocated a place in the afterlife, his ghost watches the action alongside the figure of Revenge, who often seems to be literally asleep on the job. They see that back in the land of the living, Don Andrea's love interest Bel-Imperia soon transfers her affections to Horatio, son of the Spanish Knight-Marshall Hieronimo. Beset by jealousy over this attachment, Balthazar conspires with the villainous Lorenzo to kill Horatio, whose body is discovered by his father and mother. Lorenzo

[37] Nicoleta Cinpoeş "Introduction". In *Doing Kyd: Essays on The Spanish Tragedy* ed. Nicoleta Cinpoeş (Manchester University Press, 2018), 2.

employs various machinations to cover his tracks, but Bel-Imperia manages to communicate the pair's guilt to Hieronimo using a letter written in blood (another compelling instance of the body providing 'strong proof' of authenticity). At last Hieronimo, in collaboration with Bel-Imperia, induces the pair to act in a play of Hieronimo's devising, *Soliman and Perseda*. In a striking piece of metatheatre, the knives in the scene turn out to be real, and Balthazar, Lorenzo and Bel-Imperia are killed (the latter by her own hand). Revealing his son's body behind a curtain, Hieronimo explains his plot and the events which have led to this point, before declaring his refusal to speak further:

> HIERONIMO: What lesser liberty can kings afford
> Than harmless silence? Then afford it me.
> Sufficeth I may not, nor I will not, tell thee.
> KING: Fetch forth the tortures!—Traitor as thou art, I'll make thee tell
> HIERONIMO: Indeed thou mayst torment me as his wretched son
> Hath done in murdering my Horatio.
> But never shalt thou force me to reveal
> The thing which I have vowed inviolate.
> And therefore, in despite of all thy threats,
> Pleased with their deaths, and eased with their revenge,
> First take my tongue, and afterwards my heart.
> *[He bites out his tongue]*
> KING: Oh, monstrous resolution of a wretch!
> See viceroy, he hath bitten forth his tongue
> Rather than to reveal what we required.
> CASTILE: Yet he can write.
> KING: And if in this he satisfy us not,
> We will devise th'extremest kind of death
> That ever was invented for a wretch.
> *Then Hieronimo makes signs for a knife to mend his pen*
> CASTILE: Oh he would have a knife to mend his pen.
> VICEROY: Here, and advise thee that thou write the truth.
> *Hieronimo with a knife stabs the Duke [of Castile] and himself [to death]*
> (4.4.178–96 [Q1])[38]

Hieronimo's declaration, and his subsequent actions, are undoubtedly spectacular. However, they are also puzzling. Ostensibly, Hieronimo's

[38] Further in-text references are to Q1 unless otherwise specified.

autoglossotomy is a means to prevent the King from forcibly extracting information from him about the revenge plot, but Hieronimo has already given a detailed explanation of his scheme. The only point he has omitted is the identity of his accomplice, but as Bel-Imperia is already dead, this is not a particularly valuable secret to keep. Furthermore, it is obvious—indeed, explicitly stated—that Hieronimo is already 'resolved to die' by his own hand, making his tongue-biting even more redundant (4.4.160). To approach this puzzle, we thus need to consider Hieronimo's actions as linked to suicide in the way I have described above; not causally connected, but conceptually allied through the idea of an individual usurping the state's power over their body.

Here, as elsewhere in this book, not all body parts are equal, and the area of the body which is wounded is significant. As Carlosa Mazzio has explored, the tongue in early modern culture was a particularly potent and liminal object. As the organ of speech, it was, she argues

> imagined in early modern religious, rhetorical, anatomical, and literary texts as the most powerful and the most vulnerable member of 'man,' ... Because 'tongue' (like the Latin 'lingua' and the Greek 'glossia' also means 'language,' the very invocation of the word encodes a relation between word and flesh, tenor and vehicle, matter and meaning.[39]

The intimacy of this connection between tongue-as-organ and tongue-as-language was shown by early modern texts' insistence that the strongest evidence of the necessity of restraint in speech was the anatomy of the tongue: 'The tongue is placed in the middle of the mouth, and it is compassed in with lippes ... and teeth as with a double trench, to shew us, how we are to use heed and preconsideration before wee speake'.[40] The body was here viewed as a container within which speech was both generated and policed, and the tongue as its most powerful and vulnerable boundary. In his 1592 *Treatise of the Good and Evell Tongue*, for instance, John Marconville insisted that

[39] Carla Mazzio, "Sins of the Tongue in Early Modern England". *Modern Language Studies* 28, no. 3/4 (1998): 93–124. https://www.jstor.org/stable/3195467, 94.

[40] William Perkins, "A Direction for the Government of the Tongue According to Gods Worde (Cambridge: 1595)". In *The Unruly Tongue in Early Modern England: Three Treatises*, ed. by Nathalie Vienne-Guerrin (Fairleigh Dickinson University Press, 2012), 79.

> the chiefest perfection, the best propertie perteining to a man, is to moderate his tongue, for overmuch babble is ever subject to controlment, and he that warely weareth his own tongue sheweth himselfe to be wise, and he that gards his mouth and tongue, gards his own soule.[41]

There was, therefore, traffic between the tongue imagined in literal terms as an organ bounded by the mouth, and idiomatic warnings to 'hold one's tongue' which had been in circulation for several hundred years by the time of *The Spanish Tragedy*.[42] Though the origins of the phrase 'to bite one's tongue' are obscure, it was certainly in use by 1591, when Shakespeare's *Henry VI part 2* used the phrase in its modern sense of keeping quiet:

> York must sit, and fret, and bite his tongue,
> While his own lands are bargain'd for, and sold. (1.1.219–20)

Of course, to bite one's tongue and to bite *out* one's tongue were rather different prospects, which again overlapped but were not identical in their meanings. Loss of or damage to the tongue functioned as a motif for loss of agency in numerous late-sixteenth and early-seventeenth-century plays, including Shakespeare's 1592 *Titus Andronicus*, John Marston's 1601 *Antonio's Revenge*, and Thomas Middleton's 1607 *The Revenger's Tragedy*.[43] In all these cases, tongue mutilation is a punitive means of attempting to take agency away from the tongue-less individual, though with varying degrees of success. Like the gestures described in the previous chapter, tongue-biting thus literalises something which is usually expressed figuratively; cutting out somebody's tongue is a way to make sure that they 'hold their tongue'. But the act of biting *out* the tongue had additional resonance. It accessed a tradition of tongue-biting as protest, originating in classical descriptions of autoglossotomy. Mazzio rightly points out that Hieronimo's autoglossotomy deliberately recollects that of

[41] Jean de Marconville, "A Treatise of the Good and Evell Tongue (Translated by T.S. (London: 1592))". In *The Unruly Tongue in Early Modern England: Three Treatises*, ed. Nathalie Vienne-Guerrin (Fairleigh Dickinson University Press, 2012), 47.

[42] Chaucer used the idiom first in *The Tale of Melibee* (c. 1387): 'thee is bettre holde thy tonge stille than for to speke' (Geoffrey Chaucer, *The Riverside Chaucer* (Oxford University Press, 2008), l.1219).

[43] The authorship of *The Revenger's Tragedy* is a matter for debate, having previously been assigned to Cyril Tourneur.

Zeno the Stoic, who reportedly bit out his own tongue to avoid being forced to give away secrets under torture. However, this still somewhat underestimates the depth and breadth of the tongue-biting tradition.[44] In fact, accounts of historical figures biting out their own tongues recurred throughout the sixteenth and seventeenth centuries. Marconville for instance, noted that

> An example [of autoglossotomy] … is recited by Pliny, of the Philosopher Anazerxes, who bereft himself of his tunge, & teeth also, that he wold not reveal a certain secret which the tyrant Nicocrion sought violently to wrest from him. The like we reade of Zenon a philosopher, which chose rather to cut of his tongue, then to disclose his freendes councell.[45]

That text probably drew from the 1576 *A philosophicall discourse, entituled, The anatomie of the minde*, which, in a chapter marked 'of constancie' told how

> Zeno the Stoic being cruelly tormented of a King of Cypres to utter those things which the King was desiro*us* to know, at le*n*gth because he woulde not satisfie his minde, bit of his owne toung, and spit the same in the tormentors face. But the constancie of Anaxarchus was more straung, for being taken of Nicocreon, a moste cruell of all other Tyrants, and afterwarde hearing that by the commaundement of the Tyrant he should in a morter be brused and broken into peeces, said most constantly unto him in this manner: Bruse and breake this body of mine at thy pleasure O Tyrant, yet shalt thou never diminish any whit of Anaxarchus. Then the Tyrant because he could not abide his bolde speech, commaunded that his tung should be cut out of his mouth. But Anaxarchus, laughing at his madnes, thought he should never have his minde, and therefore he bit out his owne toung, and spit the same by mamcks [mamucks, i.e. pieces] uppon the tyrants face.[46]

The 1613 *treasurie of auncient and moderne times* likewise attested that Anaxarchus (also known as Anazerxes) 'bit his tongue off in the midst betweene his teeth, because he would not disclose the secret complot

[44] Mazzio, "Sins of the Tongue".

[45] de Marconville, "A Treatise of the Good and Evell Tongue", 51.

[46] Thomas Rogers, *A Philosophicall Discourse, Entituled, The Anatomie of the Minde* (John Charlewood, 1576), fols. 2r–2v.

[against Emperor Nero]; and afterward, he threw it in the Tyraunts face'.[47] These texts make clear that while it may have been particularly iconic, Zeno's tongue-biting partook in a nascent tradition rather than single-handedly instituting a new form of protest. Hieronimo's autoglossotomy thus aligns him with a heroic trope rather than an individual hero.

Famous tongue-biters tended to have features in common. Many, Anaxarchus included, were said to be philosophers and orators, and tales of tongue-biting were reserved for rebels with a good cause rather than criminals seeking to evade justice. In a 1602 translation of Plutarch's *Morals*, for instance, it was noted that Hyperides, an Athenian politician and student of Plato, possessed 'a singular name above all other oratours, for speaking before the people', but when he fell foul of the Macedonian general Antipater, 'being set upon the racke, and put to torture, he bit his tongue off with his owne teeth, because he would not discover the secrets of the city'.[48] In all these stories, tongue-biting was more performative than pragmatic. It only temporarily solves the protagonist's problem (since the tongue-biter can always be made to write their confession instead), but is nonetheless rhetorically powerful in the same way as I have shown for Portia's 'gesture' of self-stabbing. In particular, it was frequently specified that the protagonist spat their tongue *at* their tormentor, as above and in the 1578 *Thesaurus linguæ Romanæ & Britannicæ*, which told how

> A common woman of Athens, who (after Harmodius and Aristogiton had slaine Hipparchus the tyrant) being tormented in sundrye facyons, to the intent shee shoulde discover the confederates of that murder, spake not one worde, but byting in sunder hir tongue, she spit it in the face of Hippias the tyranne, who caused hir to be tormented.[49]

The deliberate juxtaposition here of 'spake not ... but [bit her tongue]' makes clear that autoglossotomy is itself a form of utterance; less articulate than speaking, but with its own peculiar force.

[47] Thomas Miles, *The Treasurie of Auncient and Moderne Times* (William Jaggard, 1613), 41.

[48] Plutarch, *The Philosophie, Commonlie Called, the Morals Written by the Learned Philosopher Plutarch of Chæronea. Translated out of Greeke into English, and Conferred with the Latine Translations and the French, by Philemon Holland of Coventrie, Doctor in Physicke.*, trans. Philemon Holland (Arnold Hatfield, 1603), 936.

[49] Thomas Cooper, *Thesaurus Linguæ Romanæ & Britannicæ* (Henry Denham, 1578), fol. Lllllll3.

The cultural script provided by stories of the tongue and its misfortunes thus linked autoglossotomy to the Stoic ideas of self-control and self-determination described above. Where those narratives placed self-wounding in general as potentially agential, stories of classical tongue-biters situated autoglossotomy in particular as a form of self-wounding which was rhetorically effective, politically active and specifically anti-tyrannical. In so doing, these texts align unexpectedly with more modern theories of 'biopower' or 'biopolitics'—that is, of contestation over human life, its conditions and continuance. Sovereignty over the body, as Foucault suggested, consists at its most basic level in the power to decide whether a person lives or dies; or as Achille Mbembe puts it, 'To exercise sovereignty is to exercise control over mortality and to define life as the deployment and manifestation of power'.[50] In this context, suicide and self-injury can both be viewed as forms of 'necroresistance' which weaponise the body in service of political rebellion. In an analysis of self-immolation as political statement in Tibet, for instance, Archana Kaku sees such acts as uniquely placed to rebel against attempts by the state to exercise biopower over its people:

> Self-destructive violence, then, is both the refusal and negation of sovereignty. Self-destruction not only mobilizes the body in service of resistance, but ... [turns] the body—that which is within reach of even the barest form of life—into resistance. These acts do not simply mimic or make visible state violence, but rather refuse the state's assignation of certain bodies as incapable of political action. In this way, the self-immolator mobilizes their body as a subject and site of resistance, and not merely an object upon which state power is exercised.[51]

The body, which may be subject to state control, may nonetheless also be a means of resistance which remains available to individuals even when economic and political capital are stripped away. Crucially, Kaku and others argue that it is not death but *pain* which underpins the political resonance of self-injury or suicide, by demanding—and getting—a response from those who witness it. This is certainly the case for Hieronimo, who

[50] Achille Mbembé, "Necropolitics" transl. by Libby Meintjes. *Public Culture*, 15, no. 1 (2003): 11–40 (12).

[51] Archana Kaku, "Burning the Body: The Bodily Politics of Tibetan Self-Immolation". *Theory & Event* 23, no. 3 (2020): 573–606, (582). See also Banu Bargu, *Starve and Immolate: The Politics of Human Weapons* (Columbia University Press, 2014).

forces a direct confrontation between the sovereignty of the King and his own self-government. The carnage of Hieronimo's revenge extends to not only individual bodies, but the body politic; he brings death into the intimate spaces of the court and slaughters diplomatically important individuals. It is little wonder, then, that the King—previously a relaxed if somewhat oblivious ruler—steps so rapidly and violently into the role of tyrant carved out for him by Hieronimo's scene-setting. It is notable that his first response to Hieronimo's bloody playlet is to demand he 'Speak, traitor' (4.4.161), despite that fact that at that point, Hieronimo has only just finished explaining and he has not yet refused to elaborate further, nor has he yet been asked a question. In attempting to assert his authority, the King reproduces perfectly the part of the 'most cruell' tyrants who populate classical autoglossotomy stories. The monarch's next demand to 'Fetch forth the tortures' (4.4.179) confirms his desire to reassert a Foucauldian sovereignty over his subject by asserting dominion over Hieronimo's body.

Compared to the shocking violence in the rest of the play, the King's unfulfilled threat to torture Hieronimo might appear insignificant. However, it is not to be overlooked, because for Elizabethan audiences this ready turn to authoritarianism touched a nerve—namely, broader concerns about government and justice. Broadly speaking, the sixteenth century saw efforts to bring justice under the auspices of the state and out of the hands of individuals. The crackdown on suicide, as detailed above, was one such effort, though attitudes changed again in the later seventeenth century. As Deborah Willis points out, revenge tragedies of the 1590s and 1600s frequently responded to this shift towards state justice, questioning the effectiveness of central government and the ethics of personal revenge.[52] Hieronimo's dual status as Knight-Marshall and grieving father puts him on both sides of this debate. However, the King's willingness to order tortures demonstrates that it is not a simple matter of ineffective state justice versus excessive private revenge. When the state acts like a private revenger, it loses any claim to moral authority, and there is little to choose between sanctioned and unsanctioned forms of punishment. Furthermore, Timothy Turner has shown that unease over the state's use of violence may have informed *The Spanish Tragedy* to a greater

[52] Deborah Willis, "'The Gnawing Vulture': Revenge, Trauma Theory, and 'Titus Andronicus'". *Shakespeare Quarterly* 53, no. 1 (2002): 21–52. https://www.jstor.org/stable/3844038, 23–24.

extent than previously recognised. The summary justice meted out throughout the play would, he argues, have recalled for an Elizabethan audience some unsettling episodes in which their Queen had attempted to use martial law to execute her detractors without trial.[53] Even more troublingly, while torture was associated with Catholicism and Spain in particular, it was also employed in England, particularly in cases of treason. 'Given the right conditions', Turner contends, 'even the diffuse Elizabethan power structure could exhibit near-despotic tendencies'.[54]

Against this backdrop, it is Hieronimo's self-wounding, rather than his suicide, which functions as symbolically equal and opposite to torture. Kaku argues that while in torture, 'pain inflicted externally and involuntarily, reveals the openness of the body, and by extension is meant to unearth the mind's violability', the pain of the self-immolator actually 'transforms their body into a site of invulnerability' by allowing the immolator to 'speak their truth'.[55] This also rings true in an early modern context, in which the logic of torture involved the assumption that truth in some sense inhered *in* the body, and as such 'the infliction of pain could … draw truth from the body just as a knife draws blood'.[56] Belief in the necessity and efficacy of torture was linked to an increasing emphasis on interior life or 'inner self' which made prominent the capacity of individuals to withhold the truth. Sixteenth-century Protestant authorities were, for instance, troubled by the practices of equivocation and *mentalis restriction* (mental reservation) which allowed persecuted Catholics to lie even under oath. Kimberley Huth describes how in response

> Pain becomes a tool of exhibition, or more aggressively, extraction. For example in the rhetorical tradition and the legal practices that it helped to shape, pain was considered to be an effective way to disable an individual's ability to lie or dissimulate … Following Aristotle, rhetoricians and judicial practitioners considered evidence given under torture to be an "inartificial proof," one not created by the art of rhetoric … By working on and through

[53] Timothy A. Turner, "Torture and Summary Justice in 'The Spanish Tragedy'". *Studies in English Literature, 1500–1900* 53, no. 2 (2013): 277–92. https://www.jstor.org/stable/24509999, 282–83.

[54] Turner, "Torture and Summary Justice", 281; Elizabeth Hanson, "Torture and Truth in Renaissance England". *Representations* 34 (1991): 53–84. https://www.jstor.org/stable/2928770.

[55] Kaku, "Burning the Body", 587.

[56] Lisa Silverman, *Tortured Subjects: Pain, Truth, and the Body in Early Modern France* (University of Chicago Press, 2001), 22.

> the body to get at what was hidden inside, pain seems capable of producing authentic information by disabling an individual's ability to "fashion" reality as he or she may have liked.[57]

Discourses about torture and about self-wounding are thus both concerned with the body as a site of information. However, where torture attempts to put somatic experience to extremes in order to forcibly access a truth perceived to be hidden 'inside' the subject, self-wounding can draw on the aforementioned stoic discourses to construe pain as constitutive of identity, and pain inscribed on the body as expressive of that identity. In Huth's terms, 'the real significance of pain is not its ability to disclose the interior but rather its ability to be manipulated according to an individual's intentions regarding his or her outward identity'.[58] By translating an emotional experience into physical pain, self-wounding transforms individual anguish into political confrontation, and imbues meaning into apparently senseless violence.

3.3 'Hieronimo is mad again'

Does the above reading hold if we accept, as the 1615 edition of *The Spanish Tragedy* suggested, that 'Hieronimo is mad again'? This is an important question given that much of the play's popularity and influence after 1600 rested on its portrayal of mental disorder. Carol Thomas Neely has shown how additions to Kyd's tragedy fleshed out and brought to the fore Hieronimo's bouts of insanity:

> In the first extant edition (Q1 ca. 1592) ... madness is only briefly represented in a couple of scenes, which blend classical motifs with early modern vocabulary of distraction. However, in the fourth extant edition (Q4, 1602), the protagonist's madness is developed further in five (inappropriately named) "additions"; these 325 new lines are not add-ons but significant textual revisions that, by transforming Hieronimo's madness, make a new play.[59]

[57] Kimberly Huth, "Figures of Pain: Suffering and Selfhood in Early Modern English Literature" (Ph.D., Wisconsin, The University of Wisconsin - Madison, 2009), 9.

[58] Huth, "Figures of Pain", 25.

[59] Carol Thomas Neely, *Distracted Subjects: Madness and Gender in Shakespeare and Early Modern Culture*. (Cornell University Press, 2004), 33.

As Neely observes, *The Spanish Tragedy* Q4 had a formative role to play in literary depictions of madness: 'the motif of tragic madness … enlarged in the additions, and circulated widely throughout the culture, created a powerful secular model of tragic madness that would be subsequently developed for many different ends'.[60] In addition, making Hieronimo 'mad' makes the play in some respects more coherent, providing a clearer logic for both the delay in enacting revenge, and the excessive, grandiose violence of the final revenging playlet.

Nonetheless, while madness is undoubtedly important in revised versions of *The Spanish Tragedy*, the link between insanity and autoglossotomy is far from clear. Self-wounding could, in certain contexts, be taken as symptomatic of mental illness, though it was much less talked about than self-killing, which was a well-known hazard of melancholy. James Newton, for instance, advertised in 1675 his residential cures for distraction and melancholy, and mentioned that one of his (now recovered) patients 'being so void of Sense and Reason as not to know any thing that was said or done unto him, he would, when his hands were at liberty, most vigorously beat himself on his breast, and tear the hair off his head'.[61] This was consistent with Thomas Tryon's assessment that frenzy could lead people to injure themselves, sometimes as a precursor to suicide:

> when the Soul-Fires are kindled in the poysonous Root, a man can without Dread and Fear lay violent hands on himself, so wonderfully great is its power, for this is madness of the highest degree, since no man (as the Apostle said) hates his own Flesh, but loveth and cherishes it, which ought well to be considered by all Jurors concerned in such cases.[62]

More commonly, however, it was said that mad people were likely to injure themselves because they were simply insensible of pain, and 'can endure the greatest Cold, Hunger, and Stripes, without any sensible harm'.[63]

[60] Neely, *Distracted Subjects*, 45.

[61] James Newton, *They That Have Any Friends Distracted or Melancholy Are Desired To Accept This* (S.N., 1675).

[62] Thomas Tryon, *A Treatise of Dreams & Visions Wherein the Causes, Natures, and Uses, of Nocturnal Representations, and the Communications Both of Good and Evil Angels, as Also Departed Souls, to Mankind Are Theosophically Unfolded* (S.N., 1689), 282. Tryon's allusion to jurors here suggests that he too was primarily concerned with suicide, but left the definition of 'violence' open to include self-injury of various kinds.

[63] David Irish, *Levamen Infirmi: Or, Cordial Counsel to the Sick and Diseased*, (Isaac Walker, 1700), 45.

Hieronimo's potential madness may fit this description, since despite what ought to be disabling pain from his mutilation, he retains the self-possession to trick his (admittedly very gullible) captors and commit another murder before killing himself. Nonetheless, Kyd refuses to indicate that his protagonist suffers from a humorally driven malady such as melancholia; such a diagnosis would be far too definitive. Instead, he posits madness as a vague and inchoate possibility to be read as the audience pleases, so that it is always unclear whether Hieronimo is affecting madness in order to further his plot or is genuinely delusional. For example, in act 3.13 of the revised play, Hieronimo appears to mistake the old man Bazulto for the ghost of his murdered son, but the brevity of this episode makes it hard to discern whether this is genuine delusion, or a pose struck to emphasise his grief and wrong-foot his enemies into believing him to be insane. This doubleness extends throughout the play, which consistently offers at least two ways of viewing the unfolding plot. In one view, the characters are naturalistically driven by their circumstances and personalities, and thus as I have suggested above, wounding oneself is a choice which may logically result from the play's events. In the other, however, some version of Fate exerts a stronger pull, and nobody—including Hieronimo—is genuinely in control of their actions.

This doubleness corresponds with the play's stylised presentation of the cycle of violence in which its characters are swept up. There is a distinctly oneiric quality to the discovery of Horatio's body in the shadowy space of the garden and to the real/unreal confusion of the final scenes, which pervades all editions of the play. In act 4.3, for example, all versions present Hieronimo as stuck in a linguistic and psychological loop, endlessly turning over the same idea in his mind:

> Behoves thee then, Hieronimo, to be revenged.
> The plot is laid of dire revenge.
> Oh then, Hieronimo, pursue revenge,
> For nothing wants but acting of revenge. (4.3.28–29)

However, the 1602 additions do even more to emphasise Hieronimo's status as a man swept along by events over which he has little control. For instance, the added scene with the painter Bazardo at 3.12 sees Hieronimo predicting his own fate even before he has decided on a course of action: 'the end is death and madness. As I am never better than when I am mad, then methinks I am a brave fellow, then I do wonders' (3.12 *1602 add.*

l.159–61). Similarly, in the revised version of the tongue-biting scene, Hieronimo speaks the added line 'Now to express the rupture of my part, / First take my tongue, and afterwards my heart' (47–48). While emphasising Hieronimo's tongue-biting as a communicative act, these words also distance the autoglossotomy from heroic antecedents such as that of Zeno and Anazerxes by stressing the 'rupture' which drives Hieronimo's violence. In so doing, they weaken Hieronimo's claim to moral legitimacy, which is based on identification with those stoic exemplars. Tongue-biting is now a reaction to personal tragedy rather than a proactive strategy to indict the political system within which that tragedy has occurred.

These apparent diminutions of Hieronimo's agency correspond with a greater sense in Q4 of the role of the fates in determining events. At the close of the playlet, Hieronimo asserts that 'I am at last revenged thoroughly / Upon whose souls may heavens be yet revenged / With greater far than these afflictions' (4.4. *1602 add. l.25–26*). The King and Castile respond with invocations to 'Fall, heaven' and 'Roll all the world within thy pitchy cloud' (4.4. *1602 add. l.44,45*). Madness or potential madness here is less a psychological or physiological outcome than one of the means through which Revenge as a supernatural entity attains its ends. Moreover, the play's sophisticated 'nested' structure allows for uncertainty and ambiguity over the degree of madness and of predetermination.

We might consider this a kind of literary *anamorphosis*; that is, the deliberate setting-up of a scene to yield two or more pictures at once.[64] Anamorphic paintings are those which show a different image when viewed from one side or with a mirror, so that as Anna Riehl explains, two images may 'share the same material elements even as their meanings do not coincide'.[65] Holbein's *The Ambassadors* is a good example, in which a blur at the bottom of the painting becomes a skull when viewed from one side. An anamorphic literary work might do something similar; Riehl finds Donne's *The Extasie* to employ an 'anamorphic logic' which allows contradictory lines of argument and address to multiple audiences to coexist

[64] I use the word 'anamorphosis' here deliberately in preference to 'polysemy', which is often used to describe words or phrases with more than one possible meaning. Differing interpretations of Hieronimo's actions depend not only on single phrases but on the stance one takes towards his character. Just as anamorphosis in the visual arts allows different images to be seen in the same painting, these interpretations radically alter the tone of the play. On anamorphosis in literature, see Anna Riehl "Eying the Thought Awry: The Anamorphosis of John Donne's Poetry". *English Literary Renaissance* 39, no. 1 (2009): 141–62.

[65] Riehl, "Eying the Thought Awry", 143.

in the same poem. Though the routes through the poem are distinct, they inform one another, just as once a painting's optical illusion is revealed, that memory sticks even when viewing the image again from the original angle. Similarly, *The Spanish Tragedy* conjures a trick of the light in which Hieronimo can be both free and un-free when viewed from different perspectives because the on-stage drama contains three frames: *Soliman and Perseda* (1), the Spanish court (2) and the realm of Revenge and his bosses Pluto and Proserpine (3). Hieronimo's tongue-biting still asserts individual freedom and achieves specific ideological aims within frame two of the play-world, the court. However, his pursuit of those freedoms and aims is determined by larger forces working in frame three. As Musica suggests,

> if Nemesis and Revenge are to be considered avatars of the same power, then Revenge really *is* the author of *The Spanish Tragedy* in the sense that Andrea's death in that battle is the *sine qua non* of the entertainment for Pluto and Proserpine, *as well as* for Andrea's Ghost (and for us, the audience).[66]

Critics have long noted the metatheatricality both of the *The Spanish Tragedy*'s nested structure and its arrangement of critical moments as quasi-tableaux. Bate, for instance, observes how 'the play is thoroughly suffused with an interest in the resources and semiotic potentialities of theatre (there is, for instance, a profusion of wittily deployed props: box, dagger, rope, book, napkin, pen)'.[67] Episodes of violence are aesthetic events, described by Hattaway as having their 'origins in the pageantry of the court, the street dramas of executions'.[68] Hence, Lorenzo and Balthazar hang Horatio in the bower, but they also stab him 'thus, and thus; these are the fruits of love' (2.4.54). Like Hieronimo's tongue-biting, this is violence without a clear cause—we can assume Horatio is already dead from being hanged, so the stabs are performative rather than pragmatic. It recollects the real-life excessiveness of hanging, drawing and quartering, or trials which accused the defendant of having 'killed,

[66] Eughenii Musica, "Vindicating Revenge," in *Doing Kyd: Essays on The Spanish Tragedy*, ed. Nicoleta Cinpoeş (Manchester University Press, 2018), 49.

[67] Jonathan Bate, "Enacting Revenge: The Mingled Yarn of Elizabethan Tragedy", in *Doing Kyd: Essays on The Spanish Tragedy*, ed. Nicoleta Cinpoeş (Manchester University Press, 2018), 35.

[68] Michael Hattaway, "The Spanish Tragedy: Architectonic Design", in *Elizabethan Popular Theatre: Plays in Performance* (Routledge, 2005), 111.

poisoned, and murthered' their victim.[69] The aesthetics of violence are also at work when Hieronimo describes how he shall conclude his play 'with a strange and wondrous show ... behind a curtain' (4.1.177–78). When he reveals Horatio's corpse, his first words are 'See here my show, look on this spectacle' (4.4.88). In Hattaway's analysis of the play as a 'gestic' piece structured by memorable scenes rather than plot per se, Hieronimo's tongue biting is strangely neglected—in fact he says that 'motive-hunting' for that act is 'pointless'.[70] However, it is clear that as a theatrical device, Hieronimo's self-inflicted violence is a literal play on the 'dumb show' which is also recollected in the noiseless hanging of Horatio in the bower. Indeed, Hieronimo himself imagines the possibility of a play without words when in the added scene 3.12 he asks the artist Bazardo to 'paint a doleful cry' (3.12 *1602 add. l.123–24*). The final spectacle of autoglossotomy thus repeats the sensationalism which has been attached to violence throughout the play. Hieronimo stages his own mutilation for maximum dramatic flourish. However, if we doubt his capacity—because of insanity, or the workings of fate, or both—then the status of that act shifts from 'spectacular self-determination' to simply 'spectacle'. It may have less in common with Portia's 'strong proof' than Tamburlaine's 'pitiful sight'.

To an extent, this moment of dramatic indeterminacy reflected a criticism which had long been levelled at Stoicism. Augustinian opponents of neo-Stoicism argued that in reality, no-one could truly attain the level of imperturbability it seemed to require. Real human beings needed God's grace to even attempt to steer a virtuous moral course, and to suggest that any individual could achieve virtue without this was downright offensive. Thus, while many praised Cato as the ultimate Roman Stoic, and his suicide as an exemplar of mastery over the body, others 'present[ed] a Cato unable to govern his anger and his pride, a stoic who fails to master his passions. On this view his suicide is ascribed to shame or cowardice'.[71]

[69] Francis Bacon, *A True and Historical Relation of the Poysoning of Sir Thomas Overbury* (T.M. & A.C. for John Benson and John Playford, 1651), 18. This was an account of the infamous Overbury trial. The accused, James Franklin, was found guilty of having supplied poisons to court beauty Frances Carr, which she had slipped into Thomas Overbury's food while he was imprisoned in the Tower of London. The trial is detailed in Somerset, *Unnatural Murder*.

[70] Hattaway, "The Spanish Tragedy", 127.

[71] Catherine Edwards, "Modelling Roman Suicide? The Afterlife of Cato". *Economy and Society* 34, no. 2 (2005): 200–22. https://www.tandfonline.com/doi/full/10.1080/03085140500054578, 212.

This ideological conflict is exemplified in Montaigne's various writings on Stoicism.[72] He broadly praised Stoic ideals in his 'Of Constancy', and in 'On Judging Someone Else's Death'. However, in the later essay 'On the Inconstancy of our Actions', he admitted that human beings were subject to so many inconsistencies and vicissitudes that it was impossible to find an individual who exhibited even one moral virtue with perfect consistency.[73] Moreover, in 'Of Drunkeness' he went further to argue that tales of men such as Anaxarchus cheerfully enduring torture betrayed 'frenzy' and 'fury' as much as virtue. Faced with such feats, he concluded, 'who would not judge them to be prankes of a courage remooved from his wonted seat?'[74] Montaigne's question raised a pertinent question for representations of self-wounding. Though a feat of endurance or a self-injurious act might *seem* Stoical, or even mimic Stoic exemplars, it was difficult to know for sure if it was motivated by virtue, madness or some other force. This in turn raised an ethical issue for Kyd's audience. Do we, and should we, enjoy the sight of a man doing himself harm?

Ultimately, this question remains unresolved. Kyd's mastery of structure allows him to create a prismatic world in which two things can be true at once. Viewed from one facet, Hieronimo's apparently nonsensical self-wounding *works.* Biting out his tongue is a heroic assertion of autonomy which has the additional merit of exposing the tyranny lurking in apparently genial systems of authority. From a different angle, however, this act of self-directed violence is a tragic attempt to express his pain from a man for whom self-control was always an illusion, now exposed as such by his madness.

3.4 Conclusion

The shifting and heterogenous meanings of tongue-biting in *The Spanish Tragedy* aptly demonstrate why self-wounding was so imaginatively useful for early modern dramatists. Because self-wounding was discursively 'homeless', it could be given a home in different cultural scripts; it could

[72] See Strier, who argues that Montaigne repeatedly emphasised his personal idiosyncrasies and desires in a way inconsistent with Stoic ideals (Strier, *The Unrepentant Renaissance*, 208–29).

[73] Michel de Montaigne, *Essays Written in French by Michael Lord of Montaigne*, transl. by John Florio (Melch. Bradwood for Edward Blount and William Barret, 1613), 183–87.

[74] Montaigne, *Essays* 192.

be made to signify in a variety of ways.[75] In certain respects, self-wounding was like suicide, and it was condemned on the same grounds that it usurped God's rights over the body. However, like suicide it also remained open to alternative interpretations, such as that provided by neo-Stoic philosophy. By viewing self-wounding as an act freighted with associations from multiple discourses—Stoic, moralistic, medical and political—we can see how *The Spanish Tragedy* keeps numerous interpretive possibilities in play. In one, tongue-biting is the tragic product of a cycle of violence instigated and kept in motion by the machinations of supernatural figures, working in part through the mental infirmity of human beings. In the other, it is a deliberate and agential means of constructing one's own character through the mastery of pain, which in the process cites Stoic antecedents and positions the tongue-biter as heroic, his opponents as tyrannical. The latter reading is more dominant in early versions of the play and the former in later quartos, but both are accessible in each version.

The uses of this anamorphosis are partly aesthetic. Kyd's play is more interesting for its ability to highlight aspects of the playgoing experience. The playing out of Hieronimo's life for a viewing audience of Revenge and Don Andrea mimics the action played on stage for the benefit of Kyd's real-life audience, while the bloody conclusion of *Soliman and Perseda* raises the spectre of how play-viewing might go horribly wrong. This play therefore invites its audience to speculate on their involvement in an entertainment which constantly threatens to exceed its bounds. In addition, however, the multiple interpretive frameworks which meet and clash in Hieronimo's autoglossotomy also reflect the moral complexity of revenge and justice. Few watching *The Spanish Tragedy* could conclude that a triple murder-suicide is the best way to seek redress for one's wrongs, but placing Hieronimo's self-wounding at the centre of this melee also centres the human need to maintain individual agency at any cost. The wounded body speaks exceptionally loudly; as Kaku suggests, when pain is ignored, 'then one solution is to simply *make pain more visible*'.[76]

[75] This phrase is borrowed from Hanson, "Torture and Truth in Renaissance England", 57.
[76] Kaku, "Burning the Body", 583. My italics.

CHAPTER 4

Healing: Violent Lament as Emotional Self-Management

*CHRISTINE: *to TAMBURLAINE* And who did you do it for?*
TAMBURLAINE: Well, I did it for myself. It made me look stronger. Like I could withstand anything. Any pain, any challenge, and any battle.
CARLOS: But what if no one else sees it as strong? Everyone else sees it as a weakness. Most of those who I reached out to only saw it as attention seeking.[1]

In the first chapters of this book, I focussed on self-wounding as a communicative practice, linked to gesture and oratory and subject to the same debates about persuasion, authenticity and citationality that attended those forms. In texts such as *The Spanish Tragedy* and *Julius Caesar*, I argue, the ability to hurt oneself proves self-mastery over the body in service of rational ideals, and the aims of self-wounding are explicitly outward-facing. By contrast, this and the next two chapters show how self-wounding might be constructed in terms which were in many respects antithetical to Stoicism, and instead centred individual emotional and physical experience. The subject of this chapter, self-wounding in lament, shows how instead of conquering the body in service of social goals, certain kinds of self-injury could be represented as a means of catharsis, enabling emotional and physical self-regulation. Moreover, these acts once again raised questions about the impact of cultural forms on emotional experience. Acts such as hair-tearing and face-scratching accrued meaning from the

[1] Quote from *I Am Worthy*, play devised by participants in the Self Injury Through Time theatre group, 2024.

A. Skuse, *Hurt Feelings*, Palgrave Studies in Literature, Science and Medicine, https://doi.org/10.1007/978-3-032-04565-2_4

lament tradition, and texts which depicted lament represented it as *both* ritualistic *and* a spontaneous, unfeigned reaction to woe, with potentially beneficial physiological effects. Violent lament thus raised important questions about the relationship between an individual and their humoral, cultural and social ecologies.[2]

The first part of this chapter briefly outlines the lament tradition inherited by early modern writers and adapted to new purposes. Lament was a gendered form of mourning designed both to help the souls of the dead, and to assist the living in the grieving process. In this process, physical expressions of grief including face scratching, hair tearing and breast beating were entwined with verbal lament for lost relatives. After the Reformation, religious lament fell out of favour in England, but rhetoricians continued to engage with lament as a literary form. Moreover, some writers advocated for lament as an emotional purgative which encouraged the healthy expression of grief, shifting the emphasis of lament from socio-political to psycho-physiological.

In the second and third parts of the chapter, I show how some canonical works of the late sixteenth and early seventeenth century reflect both on the usefulness of violent lament as an embodied practice, and on the difficulties of 'doing' sadness through such practice. In Edmund Spenser's 1590–1596 *The Faerie Queene*, self-wounding appears as an act of catharsis, in which expressing one's passions (both literally and figuratively) prevents dangerous melancholia and aids the grieving process. However, this practice is only appropriate to women, whose colder bodies 'naturally' predispose them to displays of passionate excess. When men lament, they are likely to be viewed as womanish or self-indulgent. Like Spenser's heroines, Shakespeare's lamenting characters present their hair-tearing and face-scratching as natural and necessary expressions of their grief. However, the reactions of other characters to those displays show that such practices are seen as potentially exacerbating rather than alleviating distress, and therefore as fostering emotional inauthenticity. Finally, in considering how self-injury helps 'do' sadness, I engage with the work of anthropologist Monique Scheer. I will argue that the early modern debates about violent

[2] I use 'ecology' here in the sense that Wilson and Sullivan propose: a body (or in this case, act) 'embedded in a larger world with which it transacts' (Mary Floyd-Wilson and Garrett A. Sullivan, Jr. "Introduction: Inhabiting the Body, Inhabiting the World". In *Environment and Embodiment in Early Modern England*, ed. Mary Floyd-Wilson and Garrett A. Sullivan, Jr. (Palgrave Macmillan, 2007), 2).

lament intersect in thought-provoking ways with Scheer's contention that emotions are tricky to pin down precisely because they are both something people experience and something we do—in other words, 'We *have* emotions and we *manifest* emotions'.[3]

4.1 Lament and Melancholy in Practice

> Oh, oh! Now smite the close-shorn head! Tear your cheeks with your nails! Ah me! I have fallen as a slave to a treacherous foe I hate, a monster of lawlessness, one that by his double tongue has turned against us all that once was friendly in his camp, changing this for that and that for this again. Oh weep for me, you Trojan women! Lost and ill-fated! Ah woe! a victim to a most unhappy lot![4]

When they inherited the lament tradition from classical tragedy, early modern authors inherited a mode of expressing grief that was female-centred and highly ritualised. Most importantly for this book, it was also a tradition which incorporated bodily actions and gestures—including violence against the self—as an integral part of its affective force. For many hundreds of years, dirges sung for the dead had followed rhetorical conventions including 'antithetical thought, structure, and style ... antiphonal and stichomythic exchange between mourners ... direct address or apostrophe ... chanting and wailing'.[5] Their contents were predictable yet potentially subversive, marrying expressions of woe with exhortations to violent revenge.[6] These powerful verbal effects accompanied bodily practices which were equally ritualised. As Rebecca Saunders explains,

> Much of the lamentation ... is of the order of the expressively inarticulate, comprised of sobs, moans, shrieks, and sighs, weeping and sudden intakes of breath, as well as of elements of dance: a rhythmic sway of the body, physical

[3] Monique Scheer, "Are Emotions a Kind of Practice (and is that what makes them have a history)? A Bourdieuian Approach to Understanding Emotion". *History and Theory 51*, no. 2 (2012): 193–220 (195).

[4] Euripides, "Trojan Women" in Euripides, *The Plays of Euripides*, transl. by E. P. Coleridge. Volume I (George Bell and Sons. 1891), l.280–90.

[5] Katharine Goodland, "'Why Should Calamity Be Full of Words?': The Representation of Women's Laments for the Dead in Medieval English and Shakespearean Drama" (Ph.D., United States -- Indiana, Purdue University, 1999), 26.

[6] Tanya Pollard, "What's Hecuba to Shakespeare?". *Renaissance Quarterly* 65, no. 4 (2012): 1060–93. https://www.jstor.org/stable/10.1086/669345.

> gestures of despair, an ecstatic or trance-like manner of performance. But the lamentation also entails performance elements rarely seen in dance: women tear their clothing; beat clenched fists on their chests; scratch their cheeks or breasts, sometimes until they bleed; tear out their hair; throw dirt on their heads; soil their clothing with dung, soot, ashes, or mud.[7]

In its fullest forms, therefore, classical lament united inarticulate utterances with highly stylised language, and formalised gesture with apparently uncontrollable phenomena such as tears. It was a mode of expression 'both incorporated into and transformed by tragedy', where its ritual aspects were separated from funerary rites.[8] Passing back and forth between literary renditions and actual folk use, lament persisted over a period of many centuries, if not millennia.

The result was a genre with the power to *do* things. Sometimes, this meant empowering women to air grievances which could not be heard elsewhere. Christine Perkell, for instance, posits that the laments for Hector in book 24 of the *Iliad* express moving sorrow, but also question the value of Hector's brand of heroism given its effects on those left behind.[9] Elsewhere, it involved inciting vengeance. The story of Hecuba, who blinds Polymestor and kills his sons after losing her own son and daughter, provided a model of lament turned to violent revenge; Tanya Pollard describes how 'Euripides' Hecuba uses lament, the ritual voicing of mourning for the dead, to transform her grief into violence that is depicted as both successful and justified'.[10] Laments also served a social function. Goodland suggests that they 'move[d] the individual mourners and the larger community through the stages of loss in order to contain the threat of chaos', providing 'an aesthetic channeling of intellectual, physical, and emotional responses to death'.[11] Significantly, this process could occur even when the women lamenting were paid to do so. Like skilled actors or rhetoricians, they 'put on' grief in order to elicit sympathetic sorrow from observers. Finally, Suter finds lament to have been 'an

[7] Rebecca Saunders, *Lamentation and Modernity in Literature, Philosophy, and Culture* (Springer, 2007), 61.

[8] Casey Dué, *The Captive Woman's Lament in Greek Tragedy* (University of Texas Press, 2006), 10.

[9] Christine Perkell, "Reading the Laments of Iliad 24," in *Lament: Studies in the Ancient Mediterranean and Beyond*, ed. Ann Suter (Oxford University Press, 2008), 93–117.

[10] Pollard, "What's Hecuba to Shakespeare?", 1066.

[11] Goodland, '"Why Should Calamity Be Full of Words?', 23.

essential factor in constructing the public face of the household, the power of the family of the deceased, and the right to claim a part of the estate of the deceased'.[12] In each case, body and voice were intertwined, and the centrality of self-wounding to lament's effectiveness was confirmed by the fact that when ancient authorities sought to curtail lament, they specifically outlawed lacerating of the flesh by mourners.[13]

What these age-old practices of lament could or should do in the early modern period was a vexed question. Lament's traditional role as part of the grieving process was admittedly diminished by Reformation-era disapproval of extravagant shows of sorrow for the deceased. Lament, it was feared, encouraged belief in purgatory and intercession, and undermined the conviction that deceased relatives enjoyed a better life in heaven: 'excessive and immoderate Grief is inconsistent with our Hope and Christianity'.[14] Nonetheless, in the later sixteenth century lament acquired a new, more literary significance, 'redirected to the theaters from the streets and churches'.[15] The rich prose and verse associated with lament made extensive use of rhetorical tropes such as apostrophe and prosopopoeia which English schoolboys were expected to master by imitation. Furthermore, it lent itself to emulation and variation, two of the principal ways in which budding rhetoricians of any age were advised to hone their skills. Thus, those Greek plays and poems in which lament was most prominent—including Ovid's *Heroides*, Aeschylus' *Agamemnon* and Euripides' *Andromache*, *Hecuba*, and *Trojan Women*—became required reading for the sixteenth-century man of learning.[16] *Hecuba*, for instance, 'appeared in thirty-seven individual or partial editions during the century ... It was translated into Spanish in 1533; French in 1544; and Italian in 1543, 1550, 1563, and 1592, for a total of seven vernacular editions ... far more than any other Greek play'.[17]

* * *

[12] Ann Suter, "Introduction," in *Lament: Studies in the Ancient Mediterranean and Beyond*, ed. Ann Suter (Oxford University Press, 2008), 9.

[13] Margaret Alexiou, *The Ritual Lament in Greek Tradition* (Rowman & Littlefield Publishers, 2002), 15.

[14] Richard Oliver, *A Sermon Preach'd at the Funeral of George Payne* (Arthur Bettesworth and William Webb, 1700), 17.

[15] Goodland, "Why Should Calamity Be Full of Words?", 160.

[16] Pollard, "What's Hecuba to Shakespeare?", 1064.

[17] Pollard, "What's Hecuba to Shakespeare?", 1064.

In this climate, the question of how lament worked as a verbal, gestural and bodily practice was once again up for debate, but the emphasis now shifted from social disruption to individual experience. Contemporary theorists of cognition and emotion held that disturbances to the sensitive faculty of the soul, generated via the senses or by the mind, were communicated to the heart, which in turn drew to itself humours which, when they flowed around the body, produced physiological effects. This was most evident in discussions of melancholy, a disorder in which, as Jeremy Schmidt summarises,

> what often began as a case of affective and cognitive disturbance caused by grief, fear, or sorrow effected a change in the body through which the original mental disease became more stubbornly rooted. Thus, moral philosophy and religion were thought in many circumstances to be appropriate as early interventions in cases of severe melancholy, and in some circumstances even after, when they were combined with medicine.[18]

According to this logic, melancholy was as much or more an existential state as a humoral imbalance, and people experiencing melancholy therefore had a moral duty to divert and control their thoughts, and ask themselves: 'Thou art discontent, thou art sad and heavy, but why, upon what ground?'[19] What was more, this questioning was in itself a variety of spiritual physic which was, as Schmidt argues, hardly separable from other interventions such as exercise and purging.[20] Nonetheless, people with excessive amounts of black bile in their humoral complexion were frequently afflicted with melancholy as 'a Chronick and continuate disease'.[21] Their sufferings could reach such a pitch that suicide was a real danger, as they were 'torne in pieces with suspition, feare, sorrow, discontents, cares, shames, anguish &c. as so many wild horses, that they cannot be quiet an houre, a minute of time'.[22] When a patient had thus reached 'deaths doore', they were unlikely to be reached solely through social or spiritual

[18] Jeremy Schmidt, *Melancholy and the Care of the Soul: Religion, Moral Philosophy and Madness in Early Modern England* (Routledge, 2016), 12.

[19] Burton, *The Anatomy of Melancholy*, 363.

[20] Schmidt, *Melancholy and the Care of the Soul*, 36–38.

[21] Burton, *The Anatomy of Melancholy*, 19.

[22] Burton, *The Anatomy of Melancholy*, 271–72.

means, and required all the physician's skill as well as the moralist's counsel to be restored to health.[23]

Against this backdrop, the role of lament (or of literary and theatrical representations of lament) in fostering emotional health was up for debate. In general, works which excited violent emotion in audiences were viewed with caution, and Stephen Pender argues that a tripartite strategy of 'diversion, emolument, rectification' was consistently recommended as most suitable for melancholics.[24] Art worked as therapy when it distracted the grieving person from their sorrow, provided exemplars of people bearing great tragedy without succumbing to despair, or reasoned for the futility of weeping over somebody whose sufferings were now over. Thus, texts such as the 1671 *A Consolatory Discourse* exhorted that 'Reason and Religion must be called up in all haste, to make as strong a dam as we are able to our sorrow, or else if it have its course, it may overflow us'.[25] Not only was excessive grief immoral, warned the author, but was likely to bring about the wrath of God, 'who is displeased to see us so angry and repining, and often inflicts worse things upon us than those which we causlesly make the matter of so dolefull complaints'.[26] Occasionally, violent lament practices were singled out in these discussions; Thomas Lye preached in a funeral sermon in 1660 that the Israelites' propensity to cut themselves and 'make … baldness between their eyes' was expressive of 'immoderate grief' disliked by God.[27]

[23] Burton, *The Anatomy of Melancholy*, 272.

[24] Stephen Pender, "Rhetoric, Grief, and the Imagination in Early Modern England". *Philosophy & Rhetoric* 43, no. 1 (2010): 54–85. https://www.jstor.org/stable/10.5325/philrhet.43.1.0054, 58.

[25] Simon Patrick, *A Consolatory Discourse to Prevent Immoderate Grief for the Death of Our Friends* (Printed for R.W. for Francis Tyton, 1671), 91.

[26] Patrick, *A Consolatory Discourse*, 191. The idea that God would be angered by excessive mourning, and might inflict further sorrows as a punishment, was also found in other texts, including the 1695 *An essay on* grief, which cautioned that 'the greatest Danger that our immoderate Grief at any Loss we receive, can bring upon us, is, that it will provoke Providence to inflict greater Evils upon us. God expects a chearful Submission from us to all his Appointments, and that we should patiently wait upon his Will and Pleasure' R. W., *An Essay on Grief: With the Causes and Remedies of It* (Leon. Lichfield, 1695), 197.

[27] Thomas Lye, *The King of Terrors Metamorphosis. Or, Death Transform'd into Sleep. A Sermon Preached at the Funeral of Mrs. Elizabeth Nicoll, Daughter to That Worthy, Eminently Pious, and Charitable Citizen of London, Mr. John Walter Deceased, and Late Wife of Mr. William Nicoll of London Draper* (Printed by M.S. for Hen. Cripps, 1660), 23. 'Baldness between the eyes' refers to the practice of shaving the hair above the forehead in remembrance of the dead, and both the practices to which Lye refers are from Deuteronomy 14:1

At the same time, however, it was undeniable that people *did* mourn for their losses, and often felt better for doing so. The Sunday after the Great Fire of London, a sermon preached by Nathaniel Hardy on 'lamentation, mourning and woe' advised listeners that weeping and lamentation could form part of appropriate mourning, and noted that Jesus was recorded as having wept seven times in his life.[28] Expressions of grief were complexly related to mortification, a process which, as Chap. 6 describes, worried and fascinated early modern writers in equal measure. In 1646, Edmund Gregory's text on 'Christian melancholy' insisted that Christians ought to be 'urging ourselves to tears' on a regular basis:

> we do seldom finde the true and effectual comfort of Repentance without tears: when the eye can kindly run down with streams of water, then doth our heart begin to feel ease, then doth that burden begin to be light, which before was so heavie; and then will the light of grace begin to shine in upon our souls.[29]

As Gregory's words suggest, real people did not always adhere to, or even agree with, ideals of restraint in their emotional expression. In his *The Unrepentant Renaissance*, Richard Strier asserts that, for all that emotional temperance was widely counselled, there was equally a strong vein of anti-Stoic discourse present in discussions of emotion from the beginning of the Renaissance, through the seventeenth century.[30] Rather, he argues,

('You are the children of the Lord your God: you shall not cut yourselves, nor make any baldness between your eyes for the dead').

[28] Nathaniel Hardy, *Lamentation, Mourning, and Woe Sighed Forth in a Sermon Preached in the Parish-Church of St. Martin in the Fields, on the 9th Day of September : Being the next Lords-Day after the Dismal Fire in the City of London* (Thomas Newcomb, 1666), 10.

[29] Edmund Gregory, *An Historical Anatomy of Christian Melancholy, Sympathetically Set Forth, in a Threefold State of the Soul. 1 Endued with Grace, 2 Ensnared in Sin, 3 Troubled in Conscience* (Humphrey Moseley, 1646), 90, 5.

[30] Richard Strier, *The Unrepentant Renaissance: From Petrarch to Shakespeare to Milton* (University of Chicago Press, 2011). Notably, Strier also argues that the importance of humoralism in early modern thought has been over-emphasised: 'When an early modern person got angry, for instance, he or she did not say to her/himself: "Oh dear, my liver is heating up; my choleric humour is being activated, etc." … Instead, then, as now, people said, "That makes me mad, and I'm going to try to retaliate, or remember," or something along those lines. To make a phenomenology historical, one needs to reconstruct the kinds of things that made persons at some time react (anger, in both men and women, to continue the instance, was often tied to ideas about honor). To think that people, then or now, directly experience (or experienced) their emotions in terms of scientific theories about the physio-

people of the sixteenth and seventeenth centuries might have treated their emotions as important in much the same way that we do today.

One thing that is clear from these debates is that the proper way to 'do' grief was by no means settled, and so neither was the proper way to represent suffering on stage or on the page. The matter was complicated by the fact that, while Aristotelian ethics promoted emotional continence, Aristotle's *Poetics* also recommended 'catharsis'; that is, the quality of 'bringing about through a process of pity and fear [in the events enacted] the purification of those destructive and painful acts'.[31] Though it has previously been argued that the idea of catharsis was largely absent from English literature until the vernacular translation of the *Poetics* in the seventeenth century, Pollard convincingly contends that late sixteenth-century authors, and particularly dramatists, were acutely conscious of the 'emotional transformations' which could be wrought by their work.[32] Catharsis thus linked discussions of emotion to discussions of lament; in her study of complaint, a genre closely related to that of lament, Emily Shortslef argues that 'If we look at the language of self-consolation through complaint, we will see that complainants speak of complaining as a mode of purgation, a physical as well as emotional release of dangerously 'heavy' passion'.[33] Some of those who championed lament as a literary form made a very similar argument.[34] In his 1589 *Arte of English Poesie*, for instance, George Puttenham asserted that:

> Lamenting is altogether contrary to rejoicing, every man said so, and yet is it a peece of joy to be able to lament with ease, and freely to poure forth a mans inward sorrowes and the greefs wherewith his minde is surcharged. This was a very necessary devise of the Poet and a fine, besides his poetrie to

logical bases of emotions seems to me a category mistake of a rather major kind' (16–17). Against Michael Schoenfeldt's contention that people of the early modern period prized self-control, and embedded this into their management of both body and mind, he argues that such a stance 'produces readings that are extraordinarily and consistently conservative, readings that entirely support the rule of order, reason, and restraint' (18).

[31] Aristotle, *Aristotle's Poetics: Translated and with a Commentary by George Whalley* (McGill-Queen's University Press, 1997), 69.

[32] Tanya Pollard, "Tragedy and Revenge" in *The Oxford Companion to English Renaissance Tragedy*, ed. Emma Smith and Garrett A. Sullivan Jr. (Cambridge University Press, 2010), 62.

[33] Emily Shortslef, "Weeping, Wailing, Sighing, Railing: Shakespeare and the Drama of Complaint" (PhD diss., Columbia University, 2015), 47.

[34] See Lynne Magnusson, "Shakespearean Tragedy and the Language of Lament" in *The Oxford Handbook of Shakespearean Tragedy* (Oxford University Press, 2016).

> play also the Phisitian, and not onely by applying a medicine to the ordinary sicknes of mankind, but by making the very greef it selfe (in part) cure of the disease.[35]

Pender views Puttenham's defence of lament as essentially working towards the same goals as Stoic texts, though via a different route. In this formulation, he argues, grief is circumscribed into a defined and limitable form: 'A brief and controlled period of mourning should substitute for protracted, unpredictable sorrow; Puttenham's poetic cure trumps the remedial properties of time'.[36] By giving grief a clear outlet and defined parameters, one would hopefully have a quicker return to equilibrium.

Poetic lament as it was described by Puttenham was a practice which affected the entire ecosystem of feeling; cognitive and humoral, material and immaterial. This 'ecological' approach to feeling management is reflected in the fact that 'catharsis'—a term now used exclusively in the figurative sense—also possessed a more explicitly medical sense for early modern readers, being commonly used to describe the process of literally voiding matter from the body.[37] Given this dual meaning, lament may therefore be said to be a constructive practice in multiple ways. For the lamenter, it provided an outpouring that was both figurative and literal, since violent lament entailed both the opening of the body (hair tearing, scratching) and effusion of fluids (weeping and bleeding). For audiences, lament modelled extremes of feeling in a way that might provoke onlookers to express their own griefs. The usefulness of these effects—and their potential costs—was to be represented and discussed by the foremost authors of the Elizabethan age.

[35] George Puttenham, *The Arte of English Poesie* (Richard Field, 1589), 37–38.

[36] Pender, "Rhetoric, Grief, and the Imagination in Early Modern England", 72.

[37] See for example Thomas Willis, *Pharmaceutice Rationalis: Or, An Exercitation of the Operations of Medicines in Humane Bodies* (T. Dring, C. Harper, and J. Leigh, 1679), 85; Steven Blankaart, *A Physical Dictionary* (J.D., 1684), 34. The latter describes catharsis or 'catharticum' as 'a purging Medicine, which cleanses the Stomach, the Guts, and whatsoever is vitious and heterogeneous in the Blood, and throws it into the Common-shore of the Guts'.

4.2 *The Faerie Queene*: Lament as Emotional Self-Management

In Christopher Tilmouth's analysis of Edmund Spenser's *The Faerie Queene*, he argues that a moral imperative to tame the passions resides at the poem's heart. Passions, even apparently positive ones such as pity, are traps for the unwary. They lead almost inexorably to error, and 'incontinence, if once indulged, will draw one into intemperance'.[38] Even more dangerous are those emotions described by Tilmouth as 'pageant passions', 'displays of allegorized and anatomized passions which dominate the work's set pieces'.[39] Such passions—notably, that of despair—may be presented as things which happen *to* an individual, but in reality, they are rooted in the sufferer, who must avoid indulging them. Given the opportunity, he argues, 'Passions ... will quickly consolidate their power and then overwhelm and eradicate reason, the mind's rival and proper governing force. The waging of psychomachia, a constant suppression of passionate tendencies, is therefore a moral imperative'.[40] Approaching *The Faerie Queene* with an eye to the relationship between body and soul, Michael Schoenfeldt similarly describes temperance as the 'quietly heroic' virtue which underpins the book's ethical framework.[41] Not only the mind but the body must be kept in order and balance if a person is to reach their full potential.[42] Though it may be beyond the reach of man (and certainly of women), the ultimate goal of such self-control is to attain a state in which one not only resists temptation but is serenely immune to being tempted. It seems contradictory, then, that several of Spenser's protagonists, including those who may be flawed but are ultimately striving for virtue, engage in the apparently very *in*temperate practice of violent lament. In book VI, canto 5, Serena's grief for Calapine leads her to tear out her hair and wound herself to the point of bleeding. Una's sympathy for Redcrosse in Book I, Canto 10 causes her to tear her hair and clothes, and in Book IV, Canto 8, Timias strikes himself in a fit of despair. How might these

[38] Christopher Tilmouth, *Passion's Triumph over Reason: A History of the Moral Imagination from Spenser to Rochester*, *Passion's Triumph over Reason* (Oxford University Press, 2007), 56.

[39] Tilmouth, *Passion's Triumph over Reason*, 48.

[40] Tilmouth, *Passion's Triumph over Reason*, 56.

[41] Michael Schoenfeldt, *Bodies and Selves in Early Modern England: Physiology and Inwardness in Spenser, Shakespeare, Herbert and Milton* (Cambridge University Press, 1999), 41.

[42] Schoenfeldt, *Bodies and Selves*, 42.

instances of self-wounding supplement, or problematise, Spenser's endorsement of emotional regulation?

When the virtuous Serena realises that her knightly love interest Calapine might be dead, her emotional response is violent and immediate. Serena, whose very name means peace,

> Gan teare her hayre, and all her garments rent,
> And beat her breast, and piteously her selfe torment.
> Upon the ground her selfe she fiercely threw,
> Regardlesse of her wounds, yet bleeding rise,
> That with their bloud did all the flore imbrew,
> As if her breast new launcht with murdrous knife,
> Would streight dislodge the wretched wearie life.
> There she long groveling, and deepe groning lay,
> As if her vitall powers were at strife
> With stronger death, and feared their decay,
> Such were this Ladies pangs and dolorous assay.[43]

Serena's self-wounding is situated firmly within the lament tradition; indeed, her wild distress follows on directly from the 'lament' of the wild man who informs her by signs that Calepine cannot be found. In a poem characterised by intertextuality, this hair-tearing and breast-beating thus indirectly cites the classical antecedents described above, just as I have argued that Hieronimo's tongue-biting cites Stoic exemplars. In this case, however, there is no clear ideological purpose to Serena's self-wounding. Unlike characters such as Portia, Tamburlaine and Hieronimo, whose self-injury is explicitly communicative, Serena does not seek to persuade any-one through her actions. Instead, her violent display is presented as indicative of an internal turmoil which affects the body as well as the mind and is at least partly involuntary in its expression.

Spenser emphasises the authenticity and moral legitimacy of Serena's lament by constructing it as 'natural'. As numerous scholars have discussed, early modern natural philosophy frequently perceived similarities between women and children and 'savages', all of whom were believed to be more subject to passions and less governed by reason than 'civilised'

[43] Edmund Spenser, *Spenser: The Faerie Queene*, ed. A. C. Hamilton, Shohachi Fukuda, Toshiyuki Suzuki, and Hiroshi Yamashita, 2nd ed. (Routledge, 2014), https://doi.org/10.4324/9781315834696, Book VI, Canto 5, Verse 5. Further references are given as Book; Canto; Verse.

men. At the same time, there existed the capacity for such 'naturalness' to be read in positive terms. Unspoiled by society, 'savages' in particular might be virtuous without being polite, as is shown when the wild man rescues Calepine from Turpine's discourteous assaults in Book VI canto 7. Serena's actions may be read in similar terms. She is aligned with natural forces both via her association with the wild man and by bodily connection with the forest whose floor she 'imbrews' with her blood. Accordingly, her violence is presented as being at least partly out of her rational control. It is not her mind, but her 'breast'—presumably, her heart, which has gathered to itself the powerful melancholy humours associated with her sorrowful state—which acts to 'dislodge' life by bleeding immoderately. Yet, this self-inflicted violence proves effective as a means of catharsis in both the rhetorical and medical senses, for though Serena acts 'as if' she had stabbed herself and 'as if' she is dead, she does not, in fact, die. Instead, the stanza's language of 'launching' (lancing) and 'dislodging' implies that self-injury acts here as a purgative, with the resulting swoon now resembling a crisis of the sort believed to precede recovery from feverish illness. This apparently intemperate outburst serves an emotional and physiological purpose.

Though the use of injury as purgation in this way might seem unorthodox, it has already been rehearsed in Book I by two of the poem's principal (and most virtuous) characters; the Redcrosse Knight and his love interest Una. After being tricked by the enchanter Archimago and then the witch Duessa into rejecting the pure-at-heart Una, being imprisoned by the beastly Orgoglio and then escaping only to be lured into the cave of Despair, the Redcrosse Knight is tormented by knowledge of his own sin. He is brought to the House of Holinesse, the 'allegorical core' of the book, where he undergoes a spiritual rehabilitation which includes punishing himself with 'corrosives', 'ashes and sackcloth', and fasting.[44] Redcrosse's formal therapy, like Serena's spontaneous self-regulation, is characterised by violence. The more extreme aspects of the knight's penance are evocatively described as like an 'yron whip' which inflicts bleeding wounds which are then bathed in salt water.[45] Unsurprisingly, 'his torment was so great / That like a Lyon he would cry and rore'.[46]

[44] Spenser, *Spenser: The Faerie Queene*, I.10.26.
[45] Spenser, *Spenser: The Faerie Queene*, I.10.27.
[46] Spenser, *Spenser: The Faerie Queene*, I.10.28.

It has been noted by several critics that Redcrosse's sufferings are partly martyrological in character, aligned with the 'Christian commonplace of the sin-sick soul'.[47] As Joseph Campana argues, 'These acts of purification strikingly intermingle a Catholic rhetoric of penance with the discourse of torture common to accounts of martyrdom so effectively deployed by English Protestants'.[48] I shall discuss in Chap. 6 the conceptual blurring which often took place between secular and religious representations of self-wounding. However, it is worth noting here how this ordeal exploits well-established parallels between God's painful 'medicine' and actual medical practice. Redcrosse's penance is surgical in character, with the 'superfluous flesh' of corrupted thoughts repeatedly excised 'with pincers fyrie whott / That soone in him was left no one corrupted iott'.[49] Moreover, even its less spectacular aspects also recall the medicine actually prescribed for melancholia, by bringing Redcrosse's cures into line with the advice of contemporary medical texts. Both Burton's *Anatomy of Melancholy* (1621) and Thomas Bright's earlier *Treatise of Melancholie* (1586) prescribed extensive bleeding and purges to release the 'thick' blood associated with the melancholy condition, which could otherwise wreak catastrophic psychological and physiological effects.[50] Spenser's mingling of these two discursive fields reflects the broader tendency of both medical and religious texts to figure purgation as a process in which violence was not only a side-effect but a potentially vital component.

Redcrosse's suffering brings about a corresponding anguish in his love interest Una, whose grief is a direct result of 'hearing evermore' Redcrosse's 'ruefull shriekes and gronings'. It also provides a pattern for Una's emotional pain to be likewise translated into physical harm, and for us to understand that harm as remedial. However, whereas Redcrosse's experience draws on martyrological and medical conventions, Una's hair-pulling, garment-rending reaction is more typical of the lament tradition:

[47] D. A. Beecher, "The Anatomy of Melancholy in Book I of the Faerie Queene" *Renaissance and Reformation* 24, no. 2 (1988): 85–99, (90).

[48] Joseph Campana, *The Pain of Reformation: Spenser, Vulnerability, and the Ethics of Masculinity* (Fordham University Press, 2012), 70.

[49] Spenser, *Spenser: The Faerie Queene*, I.10.26.

[50] Burton, *The Anatomy of Melancholy*, 460–70; Timothy Bright, *A Treatise of Melancholie, Containing the Causes Thereof, & Reasons of the Strange Effects It Worketh in Our Minds and Bodies: With the Physicke Cure, and Spirituall Consolation for Such As Have Thereto Adjoyned an Afflicted Conscience* (Thomas Vautrollier, 1586) 269, 275.

> … [she] often tore
> Her guiltlesse garments, and her golden heare,
> For pitty of his paine and anguish sore;
> Yet with all patience wisely she did beare,
> For well she wist, his crime could else never be clear.[51]

The motivating passion for Una's response is 'pitty', an attribute which numerous scholars of *The Faerie Queene* have identified as hazardous.[52] Pity, it is argued, can lead a well-intentioned person into grievous error and prevent the even-handed application of justice. Yet, pity here offers a point of connection between two people driven apart by deceit. Una's self-wounding shows her readiness to partake in the suffering of another, 'opening' her body in much the same way that Redcrosse's body has been opened through a punitive and purgative regimen.[53] This stands in direct contrast to Campana's assertion that Una represents a Tyndale-esque kind of Protestantism which seeks to 'utterly denude the body of its capacity for suffering or sympathy, both of which constitute idolatry'.[54] Though it may seem indulgent, Una's expressive sorrow is not incompatible with patience, or with wisdom. Rather, Una's shared suffering with Redcrosse supports Campana's larger argument that vulnerability is the real moral of Spenser's story, over and above those of holiness or temperance: 'the 1590 *Faerie Queene* proposes as its central virtue the capacity of the flesh to be receptive or vulnerable to sensation *and* to be a site of shared vulnerability that would constitute the ground of ethical behaviour'.[55] The potential for virtue to emerge from this instance of shared bodily vulnerability is visible

[51] Spenser, *Spenser: The Faerie Queene*, I.10.28.

[52] See especially Cynthia Nazarian, "Sympathy Wounds, Rivers of Blood: The Politics of Fellow Feeling in Spenser's Faerie Queene and A View of the State of Ireland". *Modern Philology* 113, no. 3 (2016): 331–52.

[53] In this respect, it functions much like weeping, a phenomenon which Thomas Dixon describes as both 'an 'expression' in the literal sense of a squeezing out' (42), and evidential. In early modern witch trials, Dixon finds, the ability or inability to weep was part of the burden of proof for or against conviction. Supposedly, real witches could not weep even when ordered to do so because their bodies were unnaturally dry (though those who did weep could still be accused of faking it) (67). Inability or unwillingness to weep was shared by other dangerous individuals including 'Stoics, Vikings, the Swiss, tigers, bears, wolves, and werewolves' (73) (Thomas Dixon, *Weeping Britannia: Portrait of a Nation in Tears* (Oxford University Press, 2015)).

[54] Joseph Campana, *The Pain of Reformation: Spenser, Vulnerability, and the Ethics of Masculinity* (Fordham University Press, 2012), 64.

[55] Campana, *The Pain of Reformation*, 17.

in the closely matched language of knight and lady—significantly, language which again straddles medicinal and moral lexicons. Redcrosse is 'recovered by wise Patience' just as Una 'with patience wisely' laments over his suffering and translates it into her own experience.[56]

Read in the context of the lament tradition, therefore, violent displays of grief, whether on one's own behalf or from pity, need not imply moral failure. Tilmouth argues that in *The Faerie Queene*, 'Man's condition is figured as an embattled one, but as a condition, too, in which characters are accorded a fair degree of self-determination, and thus the power to combat their rebellious affections through their own resources'.[57] In Tilmouth's view, these inbuilt resources are those of God-given reason; the rational self tamping down troublesome emotions. Yet, *The Faerie Queene* repeatedly stresses the interconnection which exists between thoughts or affections, passions and humours. This interconnection makes other kinds of emotional practice possible, based on reordering the body rather than (or as well as) the mind. Lament allows the expression of grief, and self-injury in lament allows that emotional expression to overlap with physical catharsis, enabling the opening of the body and the expulsion of those humours which threaten to embed melancholy as a habit of the body rather than a temporary response to tragedy. As Schoenfeldt points out, the tour of the Castle of Alma—a thinly veiled allegory for the body—in Book II shows how intensely interested Spenser was in the body as a 'dynamic and porous edifice' which needed excretion to function properly.[58] 'The ejection of noxious material', in whatever form, 'is psychologically, physiologically, and ethically necessary to Spenser's portrait of the temperate individual'.[59]

Like lament, self-wounding might have therefore been part of an emotional schema that was diametrically opposed to Stoicism, but ultimately it achieved similar results. As I argued in Chap. 3, neo-Stoicism advocated the conscious use of reason to govern one's emotional responses to outside events. This is what Tilmouth describes as the 'moral imperative' of psychomachia, with reason constantly curbing the human tendency to be buffeted by storms of feeling.[60] In the lament-as-healthy model, however,

[56] Spenser, *Spenser: The Faerie Queene*, I.10.29.
[57] Tilmouth, *Passion's Triumph over Reason*, 73.
[58] Schoenfeldt, *Bodies and Selves*, 61.
[59] Schoenfeldt, *Bodies and Selves*, 62.
[60] Tilmouth, *Passion's Triumph over Reason*, 56.

the steadying hand of rationality was replaced, or perhaps augmented, by an approach in which it was a *good* thing for the emotions to attain a pitch at which they burst out in violent displays of woe. Only then could the humoral excess associated with those emotions be purged to the point where rational responses could once again be employed.

It is important to note, however, that this positive vision of lament comes with caveats. The contrast between Una's 'natural' lament and Redcrosse's formal penance, and the similitude between Serena and the wild man, both place violent lament as a practice most appropriate for women and uncivilised or morally immature men. However, Book IV Canto 8 demonstrates the pitfalls of applying this same model to adult men. Arthur's squire Timias finds himself disconsolate after offending his beloved Belphoebe, and he expresses his sorrow in much the same way as Spenser's female lamenters. Accompanied by the 'lamentable lay' of a turtle dove, he

> forth would poure so plenteous teares
> And beat his breast unworthy of such blame,
> And knocke his head, and rend his rugged heares,
> That could have perst the hearts of Tigres & of Beares.[61]

Timias's performance of lament once again deploys bodily signs which could equally function as tactics for humoral self-management. However, his response differs in both timeliness and effectiveness. By the time Timias employs these recognisable strategies of lament, he has already spent some time in the forest, mired in melancholy:

> His wonted warlike weapons all he broke,
> And threw away, with vow to vse no more,
> Ne thenceforth euer strike in battell stroke,
> Ne euer word to speake to woman more;
> But in that wildernesse, of men forlore,
> And of the wicked world forgotten quight,
> His hard mishap in dolor to deplore,
> And wast his wretched daies in wofull plight;
> So on him selfe to wreake his follies owne despight.[62]

[61] Spenser, *Spenser: The Faerie Queene*, IV.8.4.
[62] Spenser, *Spenser: The Faerie Queene*, IV.7.39.

Wretched and wild, Timias abandons his chivalric duties, lets his hair grow long and matted, and deliberately wastes away. But his lament is morally ambiguous in comparison to the healthful laments of Una and Serena. Their violent reactions are ones which cannot be helped but which also happen *to* help, whereas his comes after some consideration; it seems to result from a pattern of behaviour which both expresses his melancholy and helps to sustain it. Their expressions of grief allow them to move on; his impedes him from continuing the knight-errantry which Lisa Celovsky identifies as crucial for the transition from youth to full 'Man-Age'.[63] Alongside his pledge never to speak to women, his woe presents a picture of arrested development and time 'wasted'. Though lament helps to remedy this state of affairs, it does so indirectly; Timias is only relieved of his distress when the turtle dove, with a jewel around her neck, flies to Belphoebe and helps reconcile the pair. Timias's weeping and wailing expresses his grief, but additional intercession from nature is required to make lament restorative.

As several scholars have pointed out, Spenser's ambivalent representation of Timias's mourning and subsequent violent lament partly reflects this character's status as an allegory for the disgraced Sir Walter Ralegh, who was likewise out of favour with his mistress, Elizabeth I.[64] In more general terms, though, the problematics of Timias's lament also reflect the way in which the moral legitimacy of violent lament rested on beliefs about the gendered nature of the emotions. Women were believed to be more emotionally labile than men by their nature. Puttenham himself argued that 'for Ladies and women to weepe and shed tears at every little greefe, it is nothing uncomely, but rather a sign of much good nature and meeknes

[63] Lisa Celovsky, "Early Modern Masculinities and The Faerie Queene". *English Literary Renaissance* 35, no. 2 (2005): 210–47, (222–23).

[64] Timias is thought by several scholars to have been an allegorical double for Sir Walter Ralegh, with his temporary estrangement from Belphoebe reflecting Ralegh's exile from courtly company after falling out with Elizabeth I. In 1592, Ralegh secretly impregnated and then married Elizabeth Throckmorton, one of the Queen's ladies-in-waiting, and was thrown into the Tower of London as a result. Spenser was apparently unimpressed by this turn of events, and Jeffrey B. Morris speculates that Timias's rather unedifying conduct in 4.7 might reflect the poet's doubts about whether Ralegh was a 'fit servant' for Elizabeth-Belphoebe (Jeffrey Morris, 'To (Re)Fashion a Gentleman: Ralegh's Disgrace in Spenser's Legend of Courtesy". *Studies in Philology* 94, no.1 (1997): 38–58 (50)). Ralegh was not restored to court until 1597, well after this part of *The Faerie Queene* was published, so Spenser's representation of Timias's return to favour is a hopeful narrative rather than a direct record of events.

of minde, a most decent propertie for that sex'.[65] Women's bodies were 'leaky', as Gail Kern Paster has shown, in consequence of their generally colder humoral makeup; they were likely to have more melancholy humours and less opportunity to 'burn off' humoral excess through vigorous physical activity.[66] Indeed, menstruation was figured as the quintessential, and most necessary, form of physical catharsis. When one combined these beliefs with the conviction that women were more 'natural' and less rational than men, it made sense that their emotional self-regulation would take place at a physiological level first and an intellectual level second. For men, whose physiology was more favourable to self-discipline, Aristotelian continence—that is, the regulation of the emotions through mental effort—was the more appropriate solution.

4.3 Shakespeare: Marks of Woe

Working at around the same time as Spenser, Shakespeare's use of lament has been far more discussed by critics. Lynne Magnusson, for instance, finds that in his early plays in particular, Shakespeare 'developed and improvised upon the grand style and his schoolroom's lessons in copious oratory as the means to "produce" and "reproduce tears and lamentations". He afterward played variations throughout his tragic career on these inventions, repeating, improvising, restraining, and varying these first experiments'.[67] Pollard sees Shakespeare's allusion to Hecuba in *Hamlet* (famous for lamenting as well as revenging) as interacting with Ophelia's woeful display, and utilising the subversive potential of lament.[68] Shortslef places him foremost among early modern dramatists who 'recuperat[e] complaining—a term typically used to signify shameful and ineffective responses to suffering—as a valuable mode of ethical and political action in the social world'.[69] Despite this interest in Shakespeare's language of lament, criticism has seldom mentioned the place of bodily practices in representations of suffering. Yet just as *Julius Caesar* attends closely to the potential for self-wounding to move and persuade,

[65] Puttenham, *The Arte of English Poesie*, 297.

[66] Gail Kern Paster, *The Body Embarrassed: Drama and the Disciplines of Shame in Early Modern England* (Cornell University Press, 1993).

[67] Magnusson, "Shakespearean Tragedy and the Language of Lament", 124.

[68] Pollard, "What's Hecuba to Shakespeare?"

[69] Shortslef, "Weeping, Wailing, Sighing, Railing: Shakespeare and the Drama of Complaint", 4.

Shakespeare's representations of hair-tearing and face-scratching in lament show a keen awareness of how embodied practices of grief might work alongside or against verbal declarations. Furthermore, the varied responses which these practices elicit help to interrogate the way in which they are both 'natural' and culturally mediated. In exploring the usefulness of violent lament as a response to sorrow, Shakespeare poses questions about how emotional experience is generated and regulated, and how it can be known by others.

In *The Rape of Lucrece*, possibly the most moving evidence of the devastating effects of Lucrece's rape occurs before the victim has had a chance to articulate her ordeal in words. While her attacker Tarquin skulks away consumed by shame, Lucrece 'desperate, with her nails her flesh doth tear' (l.739).[70] It is an act born of despair. However, as Michael Schoenfeldt has noted, it is also an act of emotional self-management, alongside verbal lament and contemplation of the story of Philomel:

> By means of her long lamentation as well as her self-laceration and her aesthetic engagement with a tableau of suffering, Lucrece actually utilizes various early modern modes of grief management. In each case, a corollary pain distracts from the original agony. And in each case, the restoration of sociality among the suffering offers limited but significant palliative effects.[71]

Lucrece's 'tearing' of her flesh is of a piece with these other tactics, and like Serena and Una's violent lament in *The Faerie Queene*, it works. Through lamenting her ordeal, she regains the strength needed to be able to name her attacker and ask that he is punished. Though her later suicide is troubling for several reasons, it is an agential and rational act rather than one undertaken in the heat of the moment.[72]

However, this is not an uncomplicated story of successful 'grief management', and Schoenfeldt's identification of self-laceration as operating *alongside* 'aesthetic engagement' is important here. As I have argued in Chap. 2, the gestural component of self-wounding makes such acts

[70] William Shakespeare, "The Rape of Lucrece", in *The Poems: Venus and Adonis, The Rape of Lucrece, The Phoenix and the Turtle, The Passionate Pilgrim, A Lover's Complaint*, 2nd ed., The New Cambridge Shakespeare (Cambridge University Press, 2006), 151–238, https://doi.org/10.1017/9780511819094.009.

[71] Schoenfeldt, *Bodies and Selves*, 199.

[72] On the apparent imperative for rape victims and self-injured women in early modern texts to subsequently kill themselves, see Chap. 6.

implicitly citational—they are effective because they are recognisably associated with certain kinds of scenarios and certain sorts of emotions. In *The Rape of Lucrece*, however, the act of citing past traumas—whether real or the product of artistic imagination—is brought to the attention of readers as an essential part of accessing and sharing emotional experience. Scholars have long dwelt on the way in which *Lucrece* probes the boundaries between metaphor and action; as Katherine Eisamann Maus argues, the poem is occupied with 'construing tropes in their strong form; that is, not as assertions of resemblance, but of identity'.[73] Lucrece is particularly adept at using existing narratives as means to articulate her own grief. She does so through invoking, and seeking to 'imitate ... well' (l.1137), the story of Philomel, the woman-nightingale who after being raped by Tereus, imitates an extreme kind of violent lament as she continually impales herself on a thorn in order to 'keep ... sharp woes waking' (l.1136). As Maus observes of this story, the pain of self-injury re-inflicts the unwelcome penetration of the original ordeal: 'The pain [of rape] demands representation, but the representation requires the experience, even the deliberate exacerbation, of pain; the relation between emotion and speech seems more a perverse reciprocity than a simple relation of cause and effect'.[74]

Notwithstanding this uncomfortable parallel, the lesson Lucrece takes from Philomel is that physical pain is a potentially productive means to make sense of trauma: 'These means, as frets upon an instrument, / Shall tune our heart-strings to true languishment' (l.1140–41). Moreover, Lucrece demonstrates that there is value in making emotional pain not only tangible, but visible. Maus argues that 'throughout the poem vision is associated with the manifest, the immediate, the unquestionably real'.[75] This is borne out by the poem's ekphrastic description of the painting of the fall of Troy, and its use by Lucrece as a way to construct her own experience. Marion Wells describes how

> Lucrece's telling connection between visual images and mourning suggests that her absorption in the representation of mourning will facilitate the passing, or at least the endurance, of her own grief. But her words also suggest

[73] Katherine Eisaman Maus, "Taking Tropes Seriously: Language and Violence in Shakespeare's Rape of Lucrece". *Shakespeare Quarterly* 37, no. 1 (1986): 66–82. https://doi.org/10.2307/2870192, 76.

[74] Maus, "Taking Tropes Seriously", 73.

[75] Maus, "Taking Tropes Seriously", 80.

> something more: the painting will provide the 'means' of mourning, the matter of grief.[76]

Lucrece, therefore, seems to intuit the usefulness of cultural scripts in doing emotional work which includes not only expressing but actually forming one's own emotional experience, and subsequently allowing her to take action both to set revenge in motion and to end a life she no longer considers tenable. This is not to say that Lucrece's self-wounding is consciously styled or inauthentic, or even planned. In her analysis of the related genre of complaint, Shortslef describes how 'Performances of complaint were at once emblematic "speaking pictures" presented to theatrical audiences as such, and actions located within the narrative frame and world of the play, and as such these performances operate at both presentational and representational levels'.[77] We may similarly understand Lucrece's self-wounding as accessing the rich lament tradition (and being understood by audiences as doing so), whilst accepting that at the level of the poem-world her actions still constitute an impulsive response to her predicament. As I will discuss below, this depiction of self-wounding as a bodily act which is potentially both spontaneous and scripted generates more questions about the nature of emotional experience, which are not easily answered.

Shakespeare's abiding interest in this theme is evidenced in his other representations of self-wounding, both poetic and dramatic, wherein the protagonists repeatedly demand that their actions be taken as 'natural' signs of emotion. In *Troilus and Cressida*, for instance, the eponymous heroine uses hair-tearing and face-scratching as means to signal the depth of her distress when she is ordered to leave Troilus and return to the Greek camp with her father:

> CRESSIDA: Tear my bright hair and scratch my praised cheeks,
> Crack my clear voice with sobs and break my heart
> With sounding Troilus. I will not go from Troy. (4.2.103–06)[78]

[76] Marion A. Wells, "'To Find a Face Where All Distress Is Stell'd': 'Enargeia', 'Ekphrasis', and Mourning in 'The Rape of Lucrece' and the 'Aeneid'". *Comparative Literature* 54, no. 2 (2002): 97–126. https://www.jstor.org/stable/4122478, 111.

[77] Shortslef, 'Weeping, Wailing, Sighing, Railing: Shakespeare and the Drama of Complaint', 21–22.

[78] William Shakespeare, *Troilus and Cressida*, ed. Anthony B. Dawson, 2nd ed., The New Cambridge Shakespeare (Cambridge University Press, 2017), https://doi.org/10.1017/9781316418062.

Self-laceration appears alongside crying as a bodily symptom, threatening a heartbreak which may be similarly literal—lists of mortality in this period quite commonly listed 'grief' as a cause of death.[79] It is unclear here whether Cressida is actually tearing her hair and face or merely promising to do so, but in either case, she seeks to use her body as evidence for her emotional state. Cressida's ready turn to these particular markers of emotional distress implies that she understands them as belonging to the well-established lament tradition, and expects others to do the same. Nonetheless, she also clearly indicates that her personal practice of lament remains genuine and spontaneous, partly because of the tangible effects it will have upon her body.

This is a reasonable assumption, given that other Shakespearean characters quite frequently allude to tearing hair as an act of woe which should be taken at face value. In *Romeo and Juliet*, for instance, Romeo complains to Friar Lawrence:

> Thou canst not speak of that thou dost not feel.
> Wert thou as young as I, Juliet thy love,
> An hour but married, Tybalt murdered,
> Doting like me, and like me banished,
> Then mightst thou speak, then mightst thou tear thy hair,
> And fall upon the ground, as I do now,
> Taking the measure of an unmade grave. (3.3.64–70)

Tearing hair and throwing oneself on the floor is, Romeo argues, an inevitable consequence of grief in one as young and passionate as himself, faced with such impossible circumstances. He is well aware that the weeping, prostrate picture of grief he presents is predictable, but this does not mean it is not genuine. Likewise. King Lear presents a recognisable but nonetheless affecting image when he 'tears his white hair' as a marker of internal conflict as he strives to out-do 'The to-and-fro-conflicting wind and rain' (3.1.11). Lear's lament at this juncture is, as Shortslef points out, a retreat from his previous refusal to weep or complain, and a discovery that his own nature is not only rational, but also 'creaturely'; that is, that his emotional ecology incorporates the humoral and the social alongside the

[79] Erin Sullivan finds that 'grief' was listed as the cause of death of at least 357 people in the Bills of Mortality for London between 1629 and 1660 (Erin Sullivan, "A Disease unto Death: Sadness in the Time of Shakespeare", in *Emotions and Health 1200–1700*, ed. Elena Carrera (Brill, 2013), 170–71).

coldly logical.[80] These characters insist that while Stoicism may be admirable, real people express emotion messily and even violently. Furthermore, as Erin Sullivan points out, Shakespeare's characters often stress the importance of expressing grief in order to maintain mental and physical health.[81] *Titus Andronicus*'s Marcus warns Titus that 'Sorrow concealed, like an oven stopped, / Doth burn the heart to cinders where it is' (2.4.34–37). Similar sentiments can be found in *Macbeth* (4.3.210–11) and *Henry VI part 2* (3.1.198–201).

Nonetheless, the very frequency with which Shakespeare invokes hair-tearing and other violent laments also demonstrates one problem with their use, which is that the balance between spontaneity and formula is a delicate one, always threatening to tip towards the latter. If too much attention is focussed on the status of actions like hair-tearing (for example) as culturally scripted, then they lose their efficacy as unfeigned and trustworthy symptoms of an interior state. In *Troilus and Cressida*, for instance, Cressida's sobbing, hair-pulling and face-scratching take on a very different aspect if the audience knows that she will soon prove unfaithful to Troilus. At best, those actions lose some of their power to move, and at worse they may look like cynical attempts at manipulation. This problem also occurs in *Much Ado About Nothing*, when Claudio reports of Beatrice that she 'down upon her knees … falls, weeps, sobs, beats her heart, tears her hair, prays, curses—'O sweet Benedick! God give me patience!' (2.3.154–56). The audience knows that this report is concocted in order to attract Benedick's attention and falsely convince him of Beatrice's affection. In choosing to recite the 'emotional script' of breast-beating and hair-tearing as markers of distress, however, Claudio shows how far those actions have become tropes which it might be possible to adopt for persuasive purposes, rather than authentic expressions of emotion. If this problem sounds familiar, it is because it encapsulates—albeit in a different form—the same conundrum which dogged discussions of acting and rhetoric, and which I discussed in Chap. 2. How can one hope to discern the genuine from the ingenuine when they look and even feel just the same? Indeed, in such circumstances, what does 'genuine' even mean?

[80] Shortslef, 'Weeping, Wailing, Sighing, Railing", 64. See also Giulio J. Pertile, "King Lear and the Uses of Mortification". *Shakespeare Quarterly* 67, no. 3 (2016): 319–43. https://doi.org/10.1353/shq.2016.0039, 337–40.

[81] Sullivan, "A Disease unto Death", 177.

It is not only the possibility of outright fakery which problematises violent lament. In *King John*, anxieties about the power of self-injury in lament to move audiences are allied to competing beliefs about the nature of emotion, against a backdrop of political machination. In act 3.4, Lady Constance is desperately worried about the fate of her son Arthur, the heir to the English throne, who has been captured his political enemies. She chooses to express her emotions by first unloosing, then pulling at her hair:

> CONSTANCE: I am not mad: this hair I tear is mine;
> My name is Constance; I was Geffrey's wife;
> Young Arthur is my son, and he is lost:
> I am not mad: I would to heaven I were!
> For then, 'tis like I should forget myself:
> O, if I could, what grief should I forget!
> Preach some philosophy to make me mad,
> And thou shalt be canonized, cardinal;
> For being not mad but sensible of grief,
> My reasonable part produces reason
> How I may be deliver'd of these woes,
> And teaches me to kill or hang myself:
> If I were mad, I should forget my son,
> Or madly think a babe of clouts were he:
> I am not mad; too well, too well I feel
> The different plague of each calamity.
> KING PHILIP: Bind up those tresses. O, what love I note
> In the fair multitude of those her hairs!
> Where but by chance a silver drop hath fallen,
> Even to that drop ten thousand wiry friends
> Do glue themselves in sociable grief,
> Like true, inseparable, faithful loves,
> Sticking together in calamity.
> CONSTANCE: To England, if you will.
> KING PHILIP: Bind up your hairs.
> CONSTANCE: Yes, that I will; and wherefore will I do it?
> I tore them from their bonds and cried aloud
> 'O that these hands could so redeem my son,
> As they have given these hairs their liberty!'
> But now I envy at their liberty,
> And will again commit them to their bonds,
> Because my poor child is a prisoner.
> [...]
> KING PHILIP: You are as fond of grief as of your child. (3.4.46–94)

The audience in this case know that though Arthur is not yet dead, Constance's grief is all too legitimate. King John has effectively ordered the murder of her child, and though Constance does not know this, later in the scene Pandulph demonstrates to the audience that for the well-informed, the young prince's death is a predictable outcome of his capture.[82] Constance's evocation of the lament context is therefore entirely appropriate. In asserting that 'I am not mad; this hair I tear is mine', Constance articulates her belief that self-inflicted violence is a legitimate expression of grief which should be recognised as such. She argues for the body-first mode of emotional self-management because she believes that the rational response to the loss of a child is suicide or madness, and thus that her choice to self-injure is, paradoxically, a choice of self-preservation. Implicitly, Constance's insistence on recognising her feelings also shows that she recognises another danger which attends great melancholy: the loss of one's ability to emote. In his work on *King Lear*, Giulio Pertile points out that under certain circumstances, mortificatory practices such as self-flagellation were understood to stimulate rather than subdue the senses, and that this quality was noted in the popular French text *Introduction à la vie devote*, authored by Saint Francis de Sales in 1609 and soon translated into numerous languages.[83] De Sales suggested that physical mortification might thus prove useful to people in whom melancholy had brought about the stupefaction of the senses.[84] This claim was repeated by Francis Bacon in his *Historie of Life and Death*, when he suggested that 'Those Things, which are used to Mortification, may bee transferred to Vivification: For both Haire Shirts, and Scourgings, and all vexations of the outward parts, doe fortifie the Attractive Force of them'.[85] As Sullivan has suggested, Constance's continued grief may therefore provide evidence that her intellective faculty remains intact.[86]

The fact that Constance employs strategies of emotional self-management does not detract from the fact that the bereaved mother nevertheless faces a serious struggle to maintain sanity and identity, in which

[82] After persuading King John's follower Hubert not to murder him, Arthur eventually dies in 4.3 from jumping out of a tower window in a doomed escape attempt.

[83] Pertile, "King Lear and the Uses of Mortification", 324.

[84] Frances of Sales, *Introduction to the Devout Life* (Rivington, 1871), 215–33.

[85] Francis Bacon, *Historie Naturall and Experimentall, of Life and Death. Or of the Prolongation of Life* (John Haviland for William Lee, and Humphrey Mosley, 1638), 307–08.

[86] Sullivan, "A Disease unto Death", 73.

she risks coming apart at the seams. Pulling her hair roots in her own body through the experience of pain and serves as a reminder that 'this hair I tear is *mine*' (my emphasis). However, the tearing of each hair 'from their bonds' also literalises the feeling of coming unmoored from oneself through grief, the dissolution which always threatens to overcome a subject engulfed by sorrow. She is, it seems, both more and less herself through the experience of grief and subsequently of self-inflicted violence. That her hair should be the particular site upon which this paradox is focussed is no coincidence. Hair is both part of the body and an addendum to it; its removal serves a range of pragmatic and ideological purposes from medical utility (head-shaving supposedly allowed the humours to transpire from the scalp more effectively, and was sometimes used as treatment for melancholy) to the shaming of criminals. It is also a gendered attribute, of particular importance to classical practices of lamentation. In scenes of women's mourning in Greek and Roman tragedy, the unloosing and pulling of women's hair is a common trope. In Andromache's lament for Hector, for instance, we are told in detail how as a prelude to singing a dirge, Andromache

> Far from off her head...cast the bright attiring thereof, the frontlet and coif and kerchief and woven band, and the veil that golden Aphrodite had given her on the day when Hector of the flashing helm led her as his bride forth from the house of Eetion, after he had brought bride-gifts past counting. And round about her came thronging her husband's sisters and his brothers' wives, who bare her up in their midst, distraught even unto death.[87]

In the classical period as in the early modern, unbinding the hair is a rejection of social normality and a playing-out on the body of the personal chaos which is felt to accompany grief. As the above passage demonstrates, however, unloosing one's hair can also carry an erotic frisson. This is in part what King Philip seeks to curtail when he repeatedly instructs Constance to 'bind up her hairs'. Lament centred the bodies and voices of women in a way that was problematic for male audiences. Moreover, the actions of the lamenting woman (hair-tearing, dancing, face-scratching, and tearing of clothes) specifically drew attention to feminine-gendered features. When Cressida, for instance, offers to tear her 'bright' hair and scratch her 'praised' cheeks, she spotlights the attractiveness of those

[87] Homer, *The Iliad*, transl. by A.T. Murray (Heinemann, 1924). Book 22, l.470–75.

features even as she threatens to mutilate them. There was therefore a feeling that these sorts of activities, while apparently spontaneous, might be deliberately used as opportunities for exhibition.

This uneasiness about the performativity of lament was linked to a more general suspicion that rather than alleviating grief, violent laments actually stirred up greater emotion. King Philip's cutting rebuke to Constance—'You are as fond of grief as of your child'—show that he too regards lament as a powerful form of emotional practice, but one which deliberately seeks the prolongation and intensification of grief rather than healthful catharsis. In his view, Constance does not need to foster feelings, since she has far too many of them already, articulated too loudly and in the wrong company. This seems unfeeling to say the least. Yet Philip's interpretation, and the audience's potential sympathy for that interpretation, are supported by the fact that Constance is shown throughout the play to be quick to anger or despair. She has been censured by Philip in 2.1 for insulting Elinor and belittling Arthur, and in 3.1 talks at length of her womanly fears and grief 'so great / That no supporter but the huge firm earth/ Can hold it up' (3.1.74–76). King Philip and Constance therefore embody competing visions of the uses of self-injury as a means to cultivate and regulate feeling. Constance sees herself using lament ingenuously, in the same way as Spenser's passionate but virtuous heroines. Guided by her emotions, she taps into this rich cultural vein as the best means available to her to purge overwhelming sorrow. By contrast, King Philip sees violent lament as a deliberate and potentially cynical practice which unhelpfully brings into being the emotion with which it is associated.

When did lament ameliorate grief, and when did it make matters worse? This debate would never be settled. In 1671, the moral commentator Simon Patrick argued that 'We may weep, but we must not wail and lament', because to do so was to cultivate greater sorrow:

> it is a degree of madness to use reason it self … by art and skill to conjure up more boisterous passions. Who would pitty him that sets his reason against himself, and studies how to be as miserable as his mind can make him? We need not be so in love with grief, as to create it to our selves.[88]

[88] Simon Patrick, *A Consolatory Discourse to Prevent Immoderate Grief for the Death of Our Friends* (R. W. for Francis Tyton, 1671), 90, 92–93.

Patrick's criticism echoed that of Shakespeare's sceptical King Philip, and recognised that, as Saunders contends, 'the language of lamentation cannot simply record catastrophe, but must also speculatively construct it'.[89] Moreover, this was not a problem confined to lament, but one which potentially affected all emotional displays. Much of the issue lies, as Scheer has identified, with the duality of emotion as something which both *happens to* and is *done by* an individual, and thus appears to have an inner and an outer quality. Yet Scheer argues that bodily manifestations of emotion such as tears and purposeful acts such as self-injury are both products of the way a person 'does' their emotional experience based on cultural and social factors:

> The use of the term 'emotional practices' should imply 1) that emotions not only *follow* from things people do, but *are* themselves a form of practice, because they are an action of a mindful body; 2) that this feeling subject is not prior to but emerges in the doing of emotion; and 3) that a definition of emotion must include the body and its functions, not in the sense of a universal, pristine, biological base, but as a locus for innate and learned capacities deeply shaped by habitual practices.[90]

In this reading, Constance's hair-tearing is of a piece with her weeping. Yet as Scheer recognises, 'ex-pressions' of grief are inevitably read as signs of what is 'really' going on inside:

> The deeply ingrained emotional practice of striving to bring avowal into congruence with 'actual feeling,' ... is what construes 'inner' feeling as 'actual,' rendering expression or avowal dependent on it. A 'performance' can by this definition never be genuine.[91]

Scheer's modern analysis articulates much of what early modern people found troubling as they considered emotions and ways of managing them. Lament is something more than (as Hamlet puts it) 'the trappings and the suits of woe' (1.2.89), but even for the lamenter it is not quite identical with 'that within which passeth show' (1.2.88). 'Doing' the emotions, as Scheer puts it, easily shades into over-doing.

[89] Saunders, *Lamentation and Modernity*, 47.
[90] Scheer, "Are Emotions a Kind of Practice?", 220.
[91] Scheer, "Are Emotions a Kind of Practice?", 215.

To this general conundrum I would finally add a reminder of the complication raised in Chap. 2. That is, when the language of lamentation includes self-injury, lament's ability to produce effects is complicated by the addition of visible wounds—which may or may not have greater authenticity and efficacy for being visible—to emotional turmoil. Physical harm both testifies to the veracity of the emotions being experienced and exteriorises the harm which may be being done by actively cultivating those emotions. In a very literal sense, then, violent lament produces the pain it purports to express, and enforces remembrance through the marks it leaves on the body.

4.4 Conclusion

It is no coincidence that two of the early modern period's most prominent authors were fascinated by the possibilities of lament. Here was a form of emotional expression which, like literature itself, was highly stylised, took place according to codified rules and yet was no less affecting for that fact. As I have argued for certain kinds of self-wounding, self-directed violence within lament was sufficiently 'readable' that it could be transplanted into different kinds of texts. In centring the female body and its emotions, however, self-wounding in lament drew particular attention to the way in which emotions were generated and regulated. Because the dispositions of the body and mind were both influenced by the humours, there was a feedback loop between performing emotions and experiencing them. In different ways, both Stoic and anti-Stoic emotional models sought to interrupt this loop through forms of emotional practice. Stoic philosophy held that by consciously resisting the urge to perform emotions, one could eventually cease to experience those emotions, at least with the same intensity as before. Conversely, the self-management model I have described above suggested that the performance of emotions relieved the psycho-physiological factors which caused them to be experienced so intensely, and thus likewise offered a return to some sort of equilibrium. The power of self-wounding in the latter model was that it gave physical shape to the process of catharsis. As the body lost tears, hair, skin or blood, it shed grief too.

However, just as we have seen that the conscious application of Stoicism encountered problems around authenticity, the therapeutic expression of emotions was similarly fraught. Emotions were intrinsically performative, and that fact meant that it was almost impossible to discern the real from the assumed. Moreover, the performance of emotion could itself generate feeling, just as the prosopopoeia of the skilled actor risked bringing about psycho-physiological change. As Saunders asserts, 'Genuineness and authenticity cannot shake off ideological interest, any more than ideology can function without constructing authentic experience'.[92] As elsewhere, the infliction of injury in lament was partly an attempt to side-step this issue, but as *King John*'s scepticism demonstrates, it was not entirely successful, for passions, sensations, affections, humours and imagination were so intertwined that 'physical proof' might be viewed as another endlessly receding target—a phenomenon which no two people could ever experience in quite the same way.

[92] Saunders, *Lamentation and Modernity in Literature, Philosophy, and Culture*, 48.

CHAPTER 5

Punishment: Self-Gelding as Social Suicide

OLLIE: Shame is an interesting word. Why do we think some people feel they have to hide their self-injury? Carlos?

CARLOS: Some self-harm attempts may be conscious, some might not. Sometimes, there's something or someone inside you that you want to rip out, so much that you'd be prepared to end your own life in order to be free of it. Other times it's impulsive and you end up in a pool of blood before you've even had the chance to think of the consequences.

OLLIE: How about you Helena?

HELENA: Well I told no one, because I held no sense for what I'd done and certainly no voice. Now I understand that self-harm was the only 'voice' I had.

Oh Wretched Woman that I am,
That I was Wed to such a Man
That should do so great Injury
Both to himself and unto me,
Mithinks, he might have let alone
The Chiefest Branch, of Natures boon.
Oh husband, husband, what have you done?[1]

[1] "THE / Quaker's Wives Lamentation / For the LOSS of / Her Husbands Jewels, / Who Gelded Himself (in Petticoat-Lane,) to Vex His Wife" (Printed for James Bissel, 1684),

A. Skuse, *Hurt Feelings*, Palgrave Studies in Literature, Science and Medicine, https://doi.org/10.1007/978-3-032-04565-2_5

In the late seventeenth century, a ballad began to circulate called 'The Quaker's Wives Lamentation for the Loss of her Husband's Jewels'. Like many ballads, it conspicuously borrowed from other texts. It was, as the broadsheet informed readers, to be sung to the tune of 'The Spinning Wheel', and the sheet featured three woodcuts which had little to do with the subject matter—one of an angry exchange in what looks like a bookseller's shop, and the others of a Quaker man and woman. In one respect, however, this ballad was more unusual. The 'jewels' of the title are not gemstones but testicles, and the Quaker's wife laments because her husband has lost these jewels by his own hand.

This ballad is one of a surprising number of early modern texts in which men were reported to have cut off their own testicles.[2] The acts described in these texts seem in many ways to fit uneasily with the self-mutilations and self-injuries we have seen thus far. They share some similarities, for like those phenomena, they are described in highly narrative terms. Wherever they appear, they follow the same structure: the circumstances leading up to the event, the event itself and what happened afterwards. They are also rich in scene-setting detail, worthy of the stage. Accounts dwell on the blood that drips through the floor and alerts residents of the room below after a man castrates himself in an upstairs chamber, or on the single stroke of the razor with which he completes the act. In other ways, however, accounts of self-gelding are profoundly unlike the other stories of self-wounding seen so far in this book. For one thing, they occur in different sorts of texts. Rather than 'high' literary forms such as poems and plays, tales of self-gelding are told primarily in medical textbooks, in books about travel and natural philosophy, and in cheap printed ballads. Accordingly, they wear their intertextuality differently. The men who wound themselves in these texts lack the strong affiliation with classical and heroic antecedents which we have seen for other sorts of self-wounding. As I will show, self-gelding still relies on cultural traditions for its intelligibility, but those traditions are the ignoble ones of public punishment and ritual humiliation.

This chapter begins by examining the significance of castration in the early modern period as a form of bodily mutilation which signified the humiliation and even the annihilation of the victim. Castration, I will

R187358. From UCSB English Broadside Ballad Archive. https://ebba.english.ucsb.edu/ballad/21317.

[2] In common with early modern texts, this chapter uses both the terms 'gelding' and 'castration' to mean removal or the testicles and scrotum. Removal of the penis was much rarer, and I specify in-text where this seems to be what is meant.

argue, was embedded in the cultural consciousness by its use as part of the torture and execution of particularly heinous criminals. It later became imagined as a non-lethal punishment, used not by the state, but as part of festive community justice which sought to shame individuals who had breached the peace through sexual misconduct. I argue that stories of *self*-gelding drew on these discourses of shame and punishment to present this form of self-injury as a reaction to various kinds of domestic discord and personal distress. Exactly who was being shamed and why was not always clear. Numerous accounts suggested that men gelded themselves as a means to punish their wives for nagging or infidelity. In others, self-gelding was represented as a reaction to sex: too much, not enough or of the wrong kind. Overall, the stories with which men explained their decision to geld themselves demonstrated the difficulty of living up to the period's masculine ideals. Yet these accounts present self-gelding as a reaction to personal shame which results in—indeed, seems to revel in—yet more shame, both private and public. Drawing on Cynthia Marshall's work on the instability of subject identity in this period, I suggest that self-gelding provides a means to explore the masochistic pleasure which may arise from abnegating masculine selfhood.

5.1 Precious Stones

In the early modern period, to lose one's 'jewels' was an unfortunate but by no means unimaginable fate. In fact, it must have seemed an omnipresent threat. Many surgical handbooks gave instructions for removing the testicles in severe cases of herniation or injury (often from shrapnel wounds suffered in battle), and it was presented as a fairly minor operation, certainly in comparison to other kinds of amputation.[3] As I have detailed elsewhere, gelding of young boys to create castrato singers was a well-established practice on the continent in the seventeenth century, and those castrati found fame in England despite the outrage their popularity

[3] M. LeClerc, *The Compleat Surgeon, or, the Whole Art of Surgery Explain'd in a Most Familiar Method*. (Printed for M. Gillyflower; T. Goodwin, and M. Wotton; J. Walthoe; and R. Parker, 1696), 238–39; Joseph de La Charriere, *A Treatise of the Operations of Surgery. Wherein Are Mechanically Explain'd the Causes of the Diseases in Which They Are Needful, … To Which Is Added, a Treatise of Wounds, … Translated from the Third Edition of the French, Enlarg'd, Corrected and Revis'd by the Author, Joseph de La Charriere*. (Printed by R. Brugis, for D. Brown, and W. Mears; and T. Ballard, 1712), 107; Daniel Turner, *The Art of Surgery* 6th ed., vol. 2 (Printed for C. Rivington, and J. Clarke, 1741), 274–77; "The Surgical Section" (1692), Wellcome Library MSMSL3/2, 211.

sometimes provoked.[4] Travel texts such as John Bulwer's *Anthropometamorphosis* described in rapt terms the eunuchs to be found in the far and middle east, where it was said that despotic but shrewd rulers created eunuchs to guard their harems, or even fattened up castrate men to eat them 'as we do capons'.[5] Theatrical audiences, who were attuned to seeing all kinds of bodily difference on stage, proved particularly fascinated by men who had lost their scrotum, penis or both. In his examination of castration anxiety in early modern culture, Gary Taylor describes how

> the noun *eunuch(s)* appears at least 240 times in at least seventy-eight different English plays written between 1580 and the closing of the theatres in 1642; early modern synonyms for the verb *castrate* appear more than 150 times in the plays of those decades. There are actual eunuchs as speaking characters in twenty-five dramatic texts written between 1600 and 1640.[6]

For all their familiarity, however, castrate men were also still objects of horror. Alexandra Shepard describes how 'In contrast to women, men were characterized as … capable of achieving a balanced and temperate constitution. This, crucially, was the bodily basis of the link between masculinity and reason which in turn provided the justification for men's claim to social and political precedence in manhood'.[7] Though that ideal body might be more imaginary than achievable, the idea of the masculine body as more bounded and complete than that of women was powerful; creating, in Gail Kern Paster's terms, 'bodily intactness as an ideological format of gender'.[8] Thus, as I have argued elsewhere, disability of all kinds was a threat to manhood.[9] The threat posed by castration was particularly acute,

[4] Alanna Skuse, *Surgery and Selfhood in Early Modern England: Altered Bodies and Contexts of Identity* (Cambridge University Press, 2021), 16–34.

[5] John Bulwer, *Anthropometamorphosis: = Man Transform'd: Or, the Artificiall Changling* (Printed by William Hunt, 1653), 357, 356.

[6] Gary Taylor, *Castration: An Abbreviated History of Western Manhood* (Routledge Ltd, 2000), 30.

[7] Alan Shepard, *Marlowe's Soldiers: Rhetorics of Masculinity in the Age of the Armada* (Ashgate Publishing, 2002), 48.

[8] Gail Kern Paster, "'In the Spirit of Men There Is No Blood': Blood as Trope of Gender in Julius Caesar". *Shakespeare Quarterly* 40, no. 3 (1989): 284–98. https://www.jstor.org/stable/2870724, 292.

[9] Alanna Skuse, "Missing Parts in The Shoemaker's Holiday". *Renaissance Drama* 45, no. 2 (2017): 161–79. https://doi.org/10.1086/694329.

because of the significance invested in the male genitals, which were, as Jacqueline Murray explains, 'the organs that symbolized [men's] natural superiority over women, with their less perfect interiorized genitals, and the organs that allowed men to dominate women and other men, through rape or the production of heirs'.[10] A man's sexual capacity, argues Murray, was central to his masculine identity, and '[c]onsequently, the locus and symbol of men's patriarchal power was also the site of their spiritual weakness and physical vulnerability'.[11] As I discuss below, men who lost their testicles found themselves facing a potential identity crisis, as they were mercilessly mocked or simply shunned by their communities.

One very tangible reason for early modern men to fear and abhor the thought of losing their testicles was the abiding association of castration with the severest kinds of interpersonal violence. For hundreds of years, removal of the testes, penis or both had been associated with degrading one's military and political enemies. Holinshed's *Chronicles*, for example, famously recorded that following the battle of Bryn Glas in 1402,

> the women of Wales cut off their privities [of the dead Englishmen], and put one part thereof into the mouthes of everie dead man, in such sort that the cullions [testicles] hoong downe to their chins; and not so contented, they did cut off their noses and thrust them into their tailes [rectums] as they laie on the ground mangled and defaced.[12]

Atrocities in warfare were nothing new, but castration was also practised in officially sanctioned forms as a deliberately frightening form of judicial punishment. Sexual crimes were particularly likely to attract such penalties; Murray, for instance, notes that in the medieval period gelding was one of the most common punishments inflicted upon men found guilty of committing homosexual acts.[13] During such events the testicles of the offender might be burned or tossed into the crowd to further their humili-

[10] Jacqueline Murray, "Sexual Mutilation and Castration Anxiety: A Medieval Perspective", in *The Boswell Thesis: Essays on Christianity, Social Tolerance, and Homosexuality* ed. Mathew Kuefler (University of Chicago Press, 2006), 265.

[11] Murray, "Sexual Mutilation and Castration Anxiety", 265.

[12] The Holinshed Project. "The Holinshed Texts (1587)". Accessed 13 January, 2017. http://english.nsms.ox.ac.uk/holinshed/texts.php?text1=1587_5006, bk. 2:528. This episode was sufficiently notorious in later years to reappear in Shakespeare's *Henry IV, Part 1,* though without explicit mention of the victims' mutilation.

[13] Murray, "Sexual Mutilation and Castration Anxiety", 254–72, 256.

ation. Indeed, she argues, for medieval people, judicial castration, like other forms of corporal punishment, was 'part of their cultural landscape' and 'an aspect of daily life'.[14]

As well as sexual transgressions, castration was held to be an appropriate response to treason. William the Conqueror, for instance, added castration and blinding to the death penalty as punishment for treason, a practice continued by his son Henry I.[15] In the sixteenth and seventeenth centuries, these sorts of penalties were less common, but Danielle Westerhof shows that as late as 1615, emasculation (removal of penis and testicles) was still inflicted on traitors as part of the bloody ritual of drawing and quartering.[16] Genital mutilation of the victim in executions served both practical and ideological purposes. A quartered, beheaded and emasculated traitor could have their body parts displayed throughout the city, or even throughout the kingdom. The scattering of their body parts provoked fears about whether they could be raised at the Last Judgement.[17] Additionally, the annihilation of the victim's body signalled the annihilation of their identity as a legitimate subject. In Westerhof's description of Hugh Despenser the Younger's execution for high treason in 1326, she describes how castration was used as part of a series of degrading events aimed at the 'systematic removal of the traitor's signs of status and self-image'.[18] This, she argues, removed the nobility which was supposed to inhere in the aristocratic body, and served 'as a means to redraw and strengthen the normative boundaries of elite identity'.[19]

[14] Murray, "Sexual Mutilation and Castration Anxiety", 260.

[15] Wånggren and Sellberg, "The Dismemberment of Will", 299.

[16] Danielle Westerhof, "Deconstructing identities on the scaffold: the execution of Hugh Despenser the Younger, 1326". *Journal of Medieval History* 33, no. 1 (2007): 87–106. https://www.tandfonline.com/doi/full/10.1016/j.jmedhist.2007.01.006, 94.

[17] Though Church authorities insisted that God could bring together a person's scattered parts for resurrection wheresoever they might be, religious and popular practices show that people remained afraid of their bodies being partitioned after death. See Skuse, *Surgery and Selfhood in Early Modern England.*

[18] A favourite of Edward II, Despenser experienced a dramatic fall from grace when Queen Isabella and Roger Mortimer took power. Found guilty of treason, he was, according to contemporary sources, hanged until unconscious, then disembowelled, decapitated, and quartered, with his genitals being removed (presumably as part of the quartering) and burned in a fire. Despenser's emasculation as part of the execution may have been connected to accusations that he was a 'sodomite', as well as to the desire to undermine his position as a member of the aristocracy (Westerhof, "Deconstructing Identities on the Scaffold", 93.)

[19] Westerhof, "Deconstructing Identities in the Scaffold", 106.

As numerous critics have argued, such extreme judicial violence, practised over centuries, made an indelible mark on the public consciousness, embedding the partition of the body as a matter of both shame and curiosity.[20] The sheer array of means by which characters could be dismembered on the Elizabethan and Jacobean stage testifies to this; audiences never seemed to tire of seeing new ways in which the body might come apart. What has been less closely examined, however, is that even as the actual use of castration as an aspect of execution fell out of use, an increasing number of texts appeared which imagined the use of gelding as a *non*-lethal corporal punishment, somewhat akin to branding, nose-cutting or ear-slitting. Perhaps the archetype for this punishment as far as early modern readers were concerned was a famous assault on the iconoclastic French religious scholar Peter Abelard (1079–1142). In the early 1100s, Abelard embarked on a passionate affair with Heloise D'Argenteuil, for whom he had been engaged as tutor. When Heloise became pregnant and the affair came to light, Heloise's uncle Fulbert, along with several henchmen, accosted Abelard in his room at night and severed his scrotum.[21] Thereafter, the story of the unfortunate lovers, told partly through their letters to one another, swiftly acquired quasi-mythological status. It gained further traction when their letters were published in a Latin translation in Paris in 1616, and numerous English translations followed. Abelard's castration may have been particularly interesting to contemporary audiences

[20] See for instance Lisa Silverman, *Tortured Subjects: Pain, Truth, and the Body in Early Modern France* (University of Chicago, 2001); Petrus Cornelis Spierenburg, *The Spectacle of Suffering: Executions and the Evolution of Repression: From a Preindustrial Metropolis to the European Experience* (Cambridge University Press, 1984); Jonathan Davies, *Aspects of Violence in Renaissance Europe* (Ashgate Publishing Ltd, 2013); Lorna Hutson, "Rethinking the 'Spectacle of the Scaffold': Juridical Epistemologies and English Revenge Tragedy". *Representations* 89, no. 1 (2005): 30–58. https://www.jstor.org/stable/10.1525/rep.2005.89.1.30.

[21] An iconoclastic religious scholar, Abelard began an affair with Heloise while they were both living in Paris in the house of Heloise's uncle, Fulbert. When Heloise became pregnant, Abelard sent her to his family in Brittany to give birth, and proposed marriage, which she reluctantly accepted. When, upon being questioned by Fulbert, Heloise denied the marriage, Abelard sent her to a convent in Argenteuil to avoid her uncle's rage. Unfortunately, Fulbert interpreted this move as an attempt by Abelard to get rid of Heloise. Abelard was accosted at night by Heloise's uncle and his henchmen, and his testicles removed in a violent attack. After the event, Abelard used his status to 'remasculinize' himself, while Heloise was obliged to remain a nun (Martin Irvine, "Abelard and (Re)Writing the Male Body: Castration, Identity, and Remasculinization". In *Becoming Male in the Middle Ages*, ed. Jeffrey Jerome Cohen and Bonnie Wheeler (Garland Publishing, 1997)).

because it seemed like an inversion of the story of the church father Origen of Alexandria (c.185–253 CE), who was said to have castrated *himself* specifically to avoid any rumours of impropriety with the female students. In turn, it fed into both joking and serious demands for the castration of lusty priests, as I demonstrate below.

Abelard's gelding was presented as tragic, but the idea his story helped to establish—that gelding could be used as a severe but survivable punishment—was reworked in the early modern period into radically different forms, many of which clustered around the theme of sexual transgression.[22] Among the most prolific of these forms was the ballad, which was growing rapidly in popularity owing to cheap printing and an increasingly urbanised population. Numerous seventeenth-century ballads described gelding as a form of mob justice meted out against men who disrupted the social order by cuckolding other men or sexually harassing women. In around 1675, for instance, a ballad called 'The Lusty Friar of Dublin' told the story of a priest who 'lodg'd at the House of a Squire, / whose Lady wou'd often presume, / To suffer this bald-pated Fryar, / to put his old Pope into Rome'. Catching the pair in the act, the squire of the story pays £100 to a local surgeon to castrate the friar. The friar, however, attempts to outsmart the squire by separately paying the surgeon to merely pretend to carry out the surgery, and instead satisfy the squire by showing him a pair of 'dog-stones'. Predictably, his attempt at double-crossing fails. When it comes to the day of the planned deception, the surgeon reneges on his promise to the friar, and really castrates him, an act which he frames as a service to the young ladies of the town:

> The Friars all running with speed,
> as soon as his Crys they did hear,
> And seeing their Brother to bleed,
> the Rogues they did tremble for fear,
> The Surgeon said, 'twas but Discretion,
> to Gueld all, were there many more,
> Then Lasses might come to Confession,
> And they not debauch them no more,
> *The Friar he cursed the Knife,*
> *The Friar he cursed the Knife;*
> *He sighs and he groans for the loss of his Stones*
> *He had rather have parted with Life.*

[22] When the perpetrators were caught, they were subjected to both castration and blinding, a penalty that reinforced the long-standing connection between injuries to the eyes and to the testes.

'The Lusty Friar' clearly incorporated elements of the Abelard story; like Abelard, the friar is staying in the home of the man whose female relative he seduces, and he likewise meets a violent comeuppance. In so doing, it capitalised on the period's virulent anti-Catholic rhetoric, which saw some commentators seriously suggest that the government ought to summarily geld any Catholic priests it could find, 'as [they do] in Sweden'.[23] However, it also participated in a ballad tradition in which the power of individuals or groups to take the law into their own hands and effect rough justice was something to be celebrated rather than censured. In the late seventeenth-century song of 'Robin the Plowman's courage', for example, the ballad opens with Harry the Miller attempting to woo Kate, who declares that she will never marry him because he is a known thief.[24] Robin the Plowman enters and starts arguing with his love rival; whereupon Kate reveals that she despises Harry because she knows him to be castrated:

> … remember the time that Bridget held,
> While Margery, Nancy, Jone, and Kate,
> did for your wicked Actions geld you,
> Therefore, therefore never stand contending,
> since I the truth have thus related,
> Where e'er you go, ye Rascal ye know,
> you are by all young women hated.

Though the miller has been castrated by a gang of women (and beaten with a ladle by 'sweet Kate'), this song makes clear that he is the blameworthy party. His castration is seen as a legitimate reaction by the community to his bad character, and it affects him as much in terms of reputation as in physical disability; we learn that 'Now for his life, he cannot get a wife'. Notably, while Robin the Plowman provides backup, it is the women in this song who run the show and dish out punishments. This partly reflected the topsy-turvy nature of the *charivari* tradition on which ballads about community justice were based. Imagining women as the instigators of violence was both a delightfully unusual turning of the tables

[23] *The Priest Gelded: Or Popery at the Last Gasp. Shewing … the Absolute Necessity of Passing a Law for the Castration of Popish Ecclesiastics* (A. M'Culloh, 1747), 13.

[24] "Robin the Plow-Man's Courage; Or, Harry the Millers Miserable Misfortune in Courting of Young Kate, Who Declared He Had Lost His Testicles, and Therefore Far Unfit for Wedlock" (Printed for P. Brooksby, J. Deacon, J. Blare, and J. Back, 1675–1696). From UCSB English Broadside Ballad Archive. http://ebba.english.ucsb.edu/ballad/21321.

and a means to add to the humiliation of the victim, who is bested by those he presumably considers his inferiors. In addition, however, there was a strong sense that when it was carried out by women, gelding provided a pleasingly direct corrective to crimes of a sexual nature—an eye for an eye, dishonour for dishonour.

This was powerfully evident in two 1680s ballads on the theme of the 'Hertfordshire Maidens', a gang of women whose mission is apparently to identify and castrate sexually transgressive men. In the first of the pair, 'The Nine Maidens Fury to the Hartford-Shire Man', a gang of woman capture and restrain a young man. They are resolved to geld him as punishment for his sexual misdemeanours:

> The reason why they would do so,
> he alwaies would be fooling
> Where ever he did come or go,
> His Courage wanted cooling:
> He was for trimming both maids & women
> they ran if they beheld him
> But now stout Doll, with Kate, and Moll,
> are all resolv'd to geld him.[25]

It is unclear from the above lines whether their victim's crimes include sexual violence or merely lascivious conduct, but the young man is on the point of losing his testicles when the gang are met by his lover, Susan. After Susan pleads with the attackers, 'Jone vow'd she'd save what Sue should have / therefore she would not geld him'. Notably, Susan's presumed ownership of her lover's body here inverts the usual rules of coverture, in which wives were the possessions of their husbands. Foiled on this occasion, the gang are apparently not put off their quest, for the subsequent 'Hartford-shire Mens Fears of the Maidens Furies' describes what is presumably the same gang of women roaming the Hertfordshire countryside looking for young men to geld:

> I Wonder that this Age is grown
> to such a vast confusion;

[25] "The Nine Maidens Fury / To The / Hartford-Shire Man. / Who Would Have Gelded Him, but That His Sweetheart Susan Coming in, Begged His Pardon". (James Bissel, 1685–1688). From UCSB English Broadside Ballad Archive. https://ebba.english.ucsb.edu/ballad/21289.

That maids won't let young-men alone,
but by a strange Intrusion
They take much pleasure to gain their treasure
their very fingers itches,
So that men's care is now to wear
a Padlock on their Breeches.[26]

The author describes how young men are afraid to go to market or to the plough in case of being beset by the maidens. This, of course, reverses the usual pattern of sexual violence in which lone females might be assumed to be vulnerable to assault. The above stanza implies a sexually predatory dimension to the maidens' activities; the padlock on men's britches protects their 'treasure', but also brings to mind a chastity belt.

Ballads were not histories. Even if they described real events, they adapted and embellished them to make a good story. Nonetheless, the recurrence of tales like these shows that there was an appetite among the reading public for imagining punitive castrations which humiliated men who disrupted community life, and especially those deemed guilty of sexual misconduct. The ballads above were clearly indebted to the charivari or skimmington ride tradition in which communities ritualistically, and sometimes violently, shamed those thought guilty of sexual bad behaviour; often men who had been cuckolded or dominated by shrewish wives.[27] As Ingram points out, those traditions in turn incorporated punishment models from official judicial contexts: 'In charivaris two contexts merged, the penal and the festive. These customs are commonly regarded as wholly independent of the formal legal system. But in fact in this period they had extremely close affinities with the shame punishments meted out officially by certain courts of law'.[28] The thread running through ballads, charivaris and judicial punishments alike was the certainty that mutilating the body was an act which reverberated through the victim's whole life. Wherever castration was represented, it was shown as a form of violence which humiliated the person, made them pariahs in their communities, and lost

[26] 'The Hartford-Shire Mens / Fears of the Maidens Furies. / It Being an Answer to the Nine Maidens Attempt in Gelding the / Young-Man' (James Bissel, 1685–1688), http://ebba.english.ucsb.edu/ballad/21290.

[27] Ingram, "Ridings, Rough Music and the 'Reform of Popular Culture' in Early Modern England". *Past & Present* 105 (1984): 79–113. https://www.jstor.org/stable/650546, 96–97.

[28] Ingram, "Ridings, Rough Music", 92.

them the respect of their loved ones. This was the emotional freight which castration as punishment would bring to bear on depictions of *self*-gelding.

5.2 Strange Justice: Self-Castration and the Patriarchal Imperative

The ballad of *The Quaker's Wife's Lamentation*, with which this chapter began, draws intriguingly on the traditions of rough justice initiated by accounts of judicial castration and popularised in ballads of punitive gelding. Like the actions of the Hertfordshire maidens, those of the Quaker man follow a clear rationale: bodily harm is a consequence for bad behaviour. However, in this as in some other self-gelding stories, it was not always clear just *whose* bad behaviour was being punished. For instance, the Quaker's refrain that 'I am lame, and thou must hault' directly echoes a common phrase from cuckold-shaming rituals such as this one from 1618, identified by Wright:

> Woe to thee, Muchael Robins,
> That ever thou wert born
> For *Blancute makes thee cuckold*
> *And thou must wear the horn*[29]

Such rituals dictate that when a women is unfaithful, her husband is to be chastised for failing to keep her honest. They serve as a reminder that while 'owning' one's partner may imply dominance, it also means being culpable for their bad choices. However, in the 'Lamentation', the punishment extends to both partners. The Quaker's wife now suffers sexual deprivation, having lost what 'I delighted more to feel, / Than e'er I did my Spinning Wheel'. In addition, however, she experiences the stigma of being presumed to have been unfaithful despite her protestations:

> Ah my Dear Husband, you're to blame,
> To bring upon us so much shame,
> For most men will both say, and swear
> That you wid go to next Horn-Fair

[29] Jonathan Wright, "The Undoing of a Man: Cuckoldry in Heywood's Edward IV". *Ben Johnson Journal* 24, no. 2 (2017): 246–67. https://www.euppublishing.com/doi/10.3366/bjj.2017.0196, 257. My emphasis.

The Quaker's wife's complaint reflects the way that in charivari, shame attached to both partners in a failing marriage. Both the individual at fault and their spouse could be ridiculed, impersonated by their neighbours, or made to parade through the streets. Indeed, there is little to distinguish this from the punitive castration ballads described above, apart from the obvious fact of who does the castrating. As in all those ballads, the root cause is sexual misconduct, and as in 'The Nine Maidens Fury', the man's partner expresses despair at losing the 'thing' she loves most dearly. A similar note was sounded in 'The Kentish Man's Cruelty, or The Miserable Maulster' [maltster]. In this tale, the problem with the protagonist's marriage is rather different, as it is the husband who finds himself unable to stay faithful. Cognisant that his constant urge to 'leap into other men's ground' is going to get him killed, he gelds himself to quell his lustiness.[30] Once again, however, the misfortune extends to his wife, who suffers reputationally and in terms of sexual satisfaction from her husband's decision: 'He did whip it quite off, and without fear or Dread, / But his wife she had rather he'd Cut off his Head'. If her husband had died, the ballad observes, she might have married again, but since he lives, she is obliged to endure both a sexless union, and the ignominy of everyone knowing about it.

The theme of self-gelding as an act with far-reaching punitive effects was not confined to ballads, but recurred in 'learned' works of history, geography or moral invective. In Charles Ancillon's *Italian Love*, a long treatise against eunuchs, he detailed how

> [Montaigne] tells us of a certain *Peasant* in his neighbourhood, that made himself an eunuch for quite different reasons, which was for meer passion and anger against his wife; this good man, as soon as he came home, was received by his wife, who was jealous of him to an extravagance, and was continually tormenting him with the usual welcome, and said any thing against him that came uppermost, and as her jealousie furnished her with malicious abuses, he made no more ado, but immediately, with his scyth that he then had in his hand, whipt off those parts which gave her so much

[30] *The Miserable Maulster: Or, The Kentish Man's Cruelty. Who, in a Passion Dismember'd Himself, to the Unspeakable Grief of his Loving Wife, and Kind Friends.* (J. Raven. N.d). From UCSB English Broadside Ballad Archive. https://ebba.english.ucsb.edu/ballad/33671.

umbrage, and without any more ceremony threw them in the good woman's face.[31]

This story provides a good example of the way in which anecdotes about self-gelding persisted over time, and might be employed in many different kinds of texts. As well as in Ancillon's work, it was also mentioned in Robert Burton's bestselling *Anatomy of Melancholy* (1621).[32] Markedly similar stories were to be found in *The Ladies' Dictionary* (1694), in which the peasant was replaced by a baker, and in *The wonder of the little world* (1676), which told how 'Johannes Fagubiensis was by nature exceeding jealous, and he found out a subtlety most worthy of a jealous Coxcomb, for he caused himself to be gelded, with this design, that if from thenceforth his Wife should conceive, she might be palpably convinced of her dishonesty'.[33] Each report shows how the festive humour of ballad accounts translated to the notionally serious business of informative works. Yet, the gleeful misogyny of such accounts could not obscure the irony underpinning these stories, which was that the men involved would have been better off addressing their domestic issues in almost any other way. Certainly, in the cases of the miserable maulster and the unfortunate Quaker, their problems might have been just as effectively—and far less painfully—solved by simply abstaining from sex. Complaints about female behaviour in these tales therefore provided an entertaining but ultimately ineffective smokescreen for the disturbing fact that men who self-gelded, even out of spite towards others, mainly punished themselves. Thus, though it is the Quaker's wife who 'laments' his loss, her husband mentions three times the 'terrible pains' he endures as a result of his decision.

[31] Charles Ancillon, *Italian Love: Or, Eunuchism Displayed. Describing All the Different Kinds of Eunuchs; Shewing the Esteem they have met with in the World, and how they came to be made so, Wherein principally is examined, whether they are capable of Marriage, and of they ought to be suffered to enter into that Holy State … occasioned by a Young Lady's falling in Love with One, who sung in the Opera at the Hay-Market, and to whom she had like to have been Married. Written by a Person of Honour* (Printed for E. Curll, 1740), 63. The similarity to stories of autoglossotomy here is striking. In both cases, hurling the disembodied organ at the antagonist serves to underline the declarative nature of the self-injury.

[32] Robert Burton, *The Anatomy of Melancholy* (John Lichfield and James Short, for Henry Cripps, 1621), 689.

[33] N. H, *The Ladies Dictionary, Being A General Entertainment of the Fair-Sex* (John Dunton, 1694), 388; Nathaniel Wanley, *The Wonder of the Little World, Or, A General History of Man* (Printed for T. Basset, R. Cheswel, J. Wriht, and T. Sawbridge, 1673), 125.

Likewise, the maulster's wife is upset by the prospect of a sexless life, but it is he who faces 'scoffs' and 'scorn' whenever he ventures outdoors.

This theme of self-gelding as punishing *oneself* was one which both ballads and medical texts repeatedly revisited, considering it from every angle. On occasion, the reasons for men's self-punishment were explicit. Ancillon's *Italian Love*, for instance, reported that a young man cut off his testicles and sent them to his mistress after finding that during an intimate encounter he was 'so unhappy as not to be master of the instruments of his passion, which would not ... obey him, but were all ice and snow, while his heart was on fire'.[34] Elsewhere, men's self-injury seemed to be precipitated by a more general sense that the proper order of things had broken down. In 1741, for instance, the physician Daniel Turner reported how

> A poor hypochondriac, or enthusiast rather, whether on a suspicion of his wife's incontinency, and thereby to discover the truth, in case she brought him any more children, which was the common rumour, or to punish his own, I could never rightly learn: but so it fell out, that in a fit of melancholy, having shut himself up in his chamber, without any apparatus, he made shorter work, with one stroak of his razor, taking both testes and scrotum away together; Upon which so great effusion of blood from the arteries ran through the floor, as made the first discovery; the people below hastening up to him, perceived him in a manner expiring by the loss of blood, and calling presently for help ... he was rescued from the most imminent danger, and the wound, after digestion, being brought into a healing condition; the poor man ashamed of his late enterprise, nor daring to stir out of the house, removed his quarters by nights, and was never after heard of.[35]

Turner's account makes particularly visible a feature shared by most medical accounts of self-gelding: their complete lack of instructional value. This story provides gory details which invest it with extra drama; the blood dripping through the floorboards was a flourish worthy of any revenge tragedy. However it remains devoid of specifics about how Turner brought the man into a 'healing condition'. This was not merely an oversight on Turner's part, since medical practitioners made no apology for including stories of self-gelding in their works more as curiosities than as useful case

[34] Ancillon, *Italian Love*, 62.

[35] Daniel Turner, *The Art of Surgery: In Which Is Laid down Such a General Idea of the Same, as Is ... Confirmed by Practice*, 6th edn, 2 vols (C. Rivington, and J. Clarke, 1741), II, 279–80.

studies. John Browne, for example, wrote in his 1678 *Discourse of Wounds* about two similar cases:

> We had two in this our City of *Norwich* which endeavoured to castrate themselves upon the very thoughts of not marrying, mistrusting that if ever they should have any children, they could not maintain them: The first of which had taken out and cut off both his Testicles, but hereby occasioning such a flux of blood as was past his skill to stop, he sendeth for a Chirurgeon of our Town who speedily stops this and heals up the wound, and cures the Patient. The second not being so couragious, but entering upon his intended operation, could not with such dexterity act his part; but upon undertaking to take the first out, he occasions such a flux of blood as he thought would speedily have rewarded his bold attempt with death. Hence [he] was forced to send for a Chirurgeon, who after having stopt the flux agglutinated the wound, and the Patient remains in very good health.[36]

Though it appears at first reading that these two men acted together, closer inspection suggests that Browne is actually reporting two incidents which have taken place in the same town in a relatively short span (this raises questions about just how many men attempted this kind of self-mutilation). It was not unusual for Browne to include other surgeons' interesting cases in his work, but these ones have little merit from a clinical point of view, since 'agglutinating' the wound might involve the application of almost any sort of ointment or dressing. Moreover, he followed this tale by indicating two more cases he had read of, both with limited medical utility:

> Benivenius … Writes of a Monk being cured, who by reason of his zeal of chastity but yet indiscreet and less rational obsequiousness did cut off his Penis; from which issued a great flux of blood, which required great care and industry to stop; For here it behoves every of the Veins to be taken up and to be bound up, that their mouths might be stopt, the which being done, the remainder of the cure was not very difficult until the wound was almost quite cured, and he healed.
>
> I have also read of a Baker who suspecting his Wife for Adultery, to be avenged of her, doth amputate his Testicles, who should she conceive, he was resolved not to own it, making himself destitute for any such operation, by his amputation, whence he should certainly know as he conjectured by

[36] John Browne, *A Compleat Discourse of Wounds, Both in General and Particular* (E. Flesher for William Jacob, 1678), 327.

> this mad trick (than which surely could never be a greater) whether his Wife was honest or not.[37]

Though the account of Benivenius's monk contained at least limited detail about the difficulty of cases in which the penis (rather than scrotum) was injured, the latter story was framed solely in terms of its psychological features. These accounts were therefore interesting for their own sake rather than any benefit they provided to the would-be surgeon's everyday practice. That fact was reinforced by the way that Browne was careful to include an important detail in the tale of the two men from Norwich. In both cases, he specifies that the men called for a surgeon when they perceived their lives to be in danger. In this way, Browne clearly demarcated self-gelding from attempted suicide, and in so doing, refocussed attention on the somewhat confused but nonetheless powerful social factors to which the men ascribed their actions.

5.3 Punishment and Pleasure?

While these stories varied widely in terms of the media in which they appeared and the specific scenarios they related, in essentials they boiled down to the same thing. The men involved found themselves stuck between a rock and a hard place, between the social imperative to fulfil certain masculine roles and the difficulty of actually doing so. As scholars including Mark Breitenberg and Alexandra Shepard have shown, prevailing patriarchal value systems held that men were superior to women on religious grounds and on the basis of their temperament and anatomy, which were closely linked. As a result of this superiority, men had a greater right than women to hold property and status, and to pass both those things on down the patrilineal line. At the higher echelons of society, this system depended, as Breitenberg has demonstrated, on men's ability to conduct relationships *through* as well as *with* women, securing homosocial bonds by the strategic use of marriage.[38]

While this might work out well on paper (for men at least), real relationships did not always follow theoretical hierarchies. Shepard has shown

[37] Browne, *A Compleat Discourse of Wounds*, 327–28.

[38] Mark Breitenberg, *Anxious Masculinity in Early Modern England* (Cambridge University Press, 1996); see also David Kuchta, *The Three-Piece Suit and Modern Masculinity: England, 1550–1850* (University of California Press, 2002), 51–70.

that despite the construction of the 'patriarchal' body as the self-contained and complete ideal from which other bodies deviated, it was painfully evident that actual masculine bodies suffered from any number of deficiencies which caused them to deviate from this ideal, including various pathologies and disabilities but also limitations as inevitable as age and class.[39] Even if it were somehow possible to perfect oneself, other people—both male and female—were another matter. As Ingram notes 'women could never be dominated to the degree implied in the patriarchal ideal', and expressed their non-compliance with this system in a vast range of ways, from nagging to infidelity.[40] In addition, men's ability to father legitimate children was undermined, as the above stories show, by factors ranging from poverty, to impotence, to other men's unwillingness to refrain from cuckolding one another. In the homosocial network which governed much of early modern economic, social and civic life, heterosexual reproduction was thus both the weakest link and the most necessary component.

This weak link may have come under particular strain during the early modern period. Tim Hitchcock and Michelle Cohen have identified 'increasing anxiety about men's sexual performance' based in part on shifting medical understandings of gender, and more widely on demographic changes which saw men marrying later in life.[41] Drawing from Lawrence Stone, Breitenberg similarly argues that the pace of social and economic change during the sixteenth and seventeenth centuries provoked increased attention to matters of gender identity and sexual ethics. The number of sexual offences brought to trial in the Elizabethan period increased dramatically, including defamation suits from women protecting their reputations.[42] There was also a 'steady stream' of new bills passed by Parliament relating to fornication, bastardy and adultery.[43] As a whole, Breitenberg contends, 'The period was decidedly nervous about social disorder in general and, at least from the perspective of men, that threat was all too easily located in misogynistic caricatures of women and in depictions of female sexuality as monstrous or destructive to themselves'.[44] In

[39] Alexandra Shepard, *Meanings of Manhood in Early Modern England* (Oxford University Press, 2006), 68.

[40] Ingram, 'Ridings, Rough Music", 97.

[41] Tim Hitchcock and Michelle Cohen "Introduction". In *English Masculinities, 1660–1800* ed. by Tim Hitchcock and Michelle Cohen (Routledge, 2014), 9.

[42] Breitenberg, *Anxious Masculinity*, 19.

[43] Breitenberg, *Anxious Masculinity*, 19.

[44] Breitenberg, *Anxious Masculinity*, 21.

her *Vernacular Bodies*, Mary Fissel provides further evidence of what she characterises as a 'crisis in paternity' in the second half of the seventeenth century (notably, the period from which all the above ballads and medical accounts are drawn).

As was the case for accounts of castration, concerns about paternity were expressed across a variety of textual genres. Ballads drew incessantly on the theme of men marrying a woman who gave birth a few months later, while medical texts pondered the question of maternal imagination and its influence on the unborn child.[45] Overall, she argues, men were more worried than ever before about the possibility that they were either being cuckolded, or raising a child not their own:

> In popular medical books, in ballads, and in political propaganda, paternity was in crisis. Medical books problematized any easy assignment of paternity, showing how easily disrupted paternal inheritance could be. In ballads, men were repeatedly deceived by women, cuckolded constantly and never sure about which children were truly theirs. On the national level, questions about monarchical succession quickly turned into scandalous rumours about the king's mistresses and the paternity of his acknowledged illegitimate sons.[46]

Men's self-gelding therefore seemed to make literal that which was otherwise expressed in figurative terms: namely, the feelings of emasculation which inevitably arose from real men's inability to live up to the patriarchal ideal. Like Portia's steadfastness or Constance's grief, men's frustration and dismay was displayed on their bodies, a form of expression both more arresting and more universal than words alone. What this does not explain, however, is why texts about this form of self-mutilation seemed to emphasise, and even revel in, the shame attached to that action, in a way that was entirely absent from discourses about self-stabbing, autoglossotomy or violent lament. As I have shown, punitive castration was associated with the deliberate humiliation of victims. Accordingly, texts about *self*-gelding, whether fictional or factual, stressed the devastating social consequences of that action. Turner's patient, for instance, found that when he had healed from his physical injuries 'the poor man ashamed of his late enterprise, nor daring to stir out of the house, removed his quarters by nights,

[45] Mary Elizabeth Fissell, *Vernacular Bodies: The Politics of Reproduction in Early Modern England* (Oxford University Press, 2004), 208–14.

[46] Fissell, *Vernacular Bodies,* 197.

and was never after heard of'.[47] Browne reserved judgement on the self-gelding patients he had personally encountered, but described the protagonists of similar incidents as 'indiscreet', irrational and 'mad'.[48] It was much the same story in ballads. For instance, when the Quaker is berated by his wife for exposing him to the censure of their neighbours at the next 'Horn-Fair', he repeatedly expresses his regret, and admits that 'I wish it were to do again / I should not do my dear such wrong'. The public ignominy potentially faced by men who gelded themselves was most insistently depicted in 'The Miserable Maulster', which devoted several verses to describing the shame the maulster faces when his injury becomes known:

Since this is a Branch which he cannot repair,
It is reason and fit that this Maulster should wear,
All the days of his life a strong Night Cap of Horn,
Thus the Women and Wenches all laugh him to scorn
  Ever since that same time such a life he has lead,
  That he wishes indeed he had Cut off his Head.

Some Women come to him and merrily joke,
Saying, prithee sweet Neighbour come play us a stroak
But the man is not able and that they all know,
And besides when along in the Town he does go,
Oh! the scoffs of the woman this Maulster does dread,
  For they tell him 'twere better he'd Cut off his head.

He once had the love of Young Women indeed,
When it lay in his power their fancies to feed,
But yet nevertheless since he's lost his good Geer,
Not a Wench of them all will his presence come near,
  But to scoff at the man; thus his comforts are fled,
  For they tell him 'twere better he'd Cut off his head.

These accounts suggest that the lasting pain of self-gelding is not physical but social, in the miserable reality of being ostracised from one's community and friends. Yet, it is clear from discourses around punitive castration that this must have been an entirely expected consequence of losing one's testicles, since both the process of castration and the fact of being castrate were occasions for censure. In other words, shame, specifically public

[47] Turner, *The Art of Surgery*, II, 279–80.
[48] Browne, *A Compleat Discourse of Wounds*, 327–28.

shaming, was not an unfortunate side effect of self-gelding, but an intrinsic part of what that act entailed. Men who gelded themselves were presented as having done so at least partly out of an urge to experience this humiliation—that is, out of a kind of masochism.

* * *

Perhaps surprisingly given the nature of the injury in question, there is little suggestion that the masochism attached to self-gelding was sexual in nature. Sexually gratifying masochistic practices such as flagellation certainly flourished during the seventeenth century, being openly discussed in works such as Robert Dixon's *Canidia*:

> The more tormented, the more he itcht,
> None can say, but he was bewitcht,
> He was conjur'd into Venus Arms
> No otherwise than by Whipping Charms[49]

As Sarah Toulalan has shown, these might be connected with punitive experiences of whipping, which 'provided a cultural framework in which scenes of flagellation for sexual pleasure may have been seen neither as unusual nor as reprehensible'.[50] However, whereas flagellation was seen as a potentially useful means to arouse sexual desire, most accounts of self-castration present the act as a means to circumvent troublesome sexual urges, and the physical pain of castration as something entirely unwanted. Instead, we might view the masochism inherent in this phenomenon in a broader sense. Several studies of early modern literature have lately identified masochistic urges at play in the depiction of various kinds of abasement. For instance, in analysing certain Shakespearean sonnets addressed to a male patron, Hugh McIntosh argues that the speaker's insistence on their lowliness in relation to the addressee's nobility constitutes a 'social masochism' which 'move[s] beyond acts of deference toward performances of self-abjection'.[51] This abjection both possesses erotic potential

[49] Robert Dixon, *Canidia, or, The Witches a Rhapsody* (1683), 6.

[50] Sarah Toulalan, *Imagining Sex: Pornography and Bodies in Seventeenth-Century England* (Oxford University Press, 2007), 116.

[51] Hugh McIntosh, "The Social Masochism of Shakespeare's Sonnets". *Studies in English Literature, 1500–1900* 50, no. 1 (2010): 109–25. https://www.jstor.org/stable/40658422, 117.

and makes visible the class structures dividing speaker from patron.[52] Similarly, John Yamamoto-Wilson points out that sadomasochistic representations of violence such as those he identifies in Nashe's *The Unfortunate Traveller* and Shakespeare's *Titus Andronicus* are often enmeshed with parody, as pain reaches heights both sublime and ridiculous. They are also, he argues, disruptive of normal ways of reading the body, and thus intrinsically linked to 'the subversion of other discourses, undermining paradigms of romantic love, martyrdom, the duty of revenge, and so on'.[53]

If masochism is identified as a powerful force in early modern culture, however, exactly what that masochism does to the concept of the self is a matter for debate. McIntosh argues that the social masochism of the sonnets brings pleasure and thus reinforces a sense of the independent, desiring self, 'an important episode in the cultural history of interiority-in-performance'.[54] In her influential *The Shattering of the Self*, however, Cynthia Marshall takes a radically different view, and argues that the Freudian concept of masochism as *self-cancelling* is particularly instructive for thinking about the early modern period. Freud, she notes, viewed masochism in his early works as an inverse kind of sadism, in which the sadist assumes the perspective of the tortured person. However, in 'Beyond the Pleasure Principle', he developed this view to argue that 'primary masochism' was not necessarily tethered to sadism, and might in fact have less to do with the sexual pleasure of self-punishment than with personal identity:

> [Previously,] clinical observations forced upon us the view that the part-instinct of masochism, the one complementary to sadism, is to be understood as a recoil of the sadism on to the ego itself. A turning of the instinct from the object to the ego is, however, essentially the same as a turning from the ego to the object, which is just now the new idea in question. Masochism, the turning of the instinct against the self, would then be in reality a return to an earlier phase of this, a regression. The exposition I then gave of masochism needs correction in one respect as being too exclusive: masochism may also be what I was there concerned to deny, primary.[55]

[52] McIntosh, 'The Social Masochism of Shakespeare's Sonnets', 123.

[53] John Yamamoto-Wilson, *Pain, Pleasure and Perversity: Discourses of Suffering in Seventeenth-Century England* (Ashgate Publishing, 2013), 189.

[54] McIntosh, 'The Social Masochism of Shakespeare's Sonnets', 123.

[55] Sigmund Freud, *Beyond the Pleasure Principle* (Dover Publications, Incorporated, 2015), 47.

In Freud's formulation, masochism is an expression of human fascination with and paradoxical desire for self-extinction—the 'death drive'. Though she does not agree with Freud in all aspects, Marshall sees this idea as applicable to early modern society, in which she argues that the dominance of a new 'individualist' mode of selfhood has been vastly overstated. While authors were undoubtedly interested in inwardness and self-cultivation, the discourses of religion and humoralism continued to depict the human condition as 'fearfully unstable' and constantly vulnerable to moral and physical influences.[56] Moreover,

> various discourses give evidence to a concept of the self in late-sixteenth and early-seventeenth-century England as not merely volatile but paradoxically affirmed in its moments of self canceling or shattering. The degree to which early modern subjects were conflicted in their emergent selfhood, not just unstable structurally but dynamically and often simultaneously pulled toward opposite extremes of dissolution and coherence, has been downplayed by humanism's developmental emphasis. In fact, the contradiction between autonomy and instability *defined* the emerging subject.[57]

In Marshall's analysis, this 'aesthetic of shattering or self-negation' helps to explain why early modern theatrical audiences were so drawn to depictions of extreme violence.[58] These images offered audiences the chance to imagine the dissolution of the coherent self, a dissolution which chimed with aspects of their everyday experience and spoke to a desire to be relieved of the burden of individual responsibility.[59]

Applying Marshall's insights into masochism and selfhood to texts about self-castration helps to explain why some men chose to injure themselves in this way specifically, and why they did so despite the shame that inevitably followed. It is notable that several ballad texts about castration, including 'The Quaker's Wife's Lamentation', 'The Miserable Maulster' and 'The Lusty Friar' are unified in their insistence that it would be better to be dead than castrate. The miserable maulster 'wishes indeed he had cut off his head', while the repeated refrain of 'The Lusty Friar' is that

[56] Cynthia Marshall, *The Shattering of the Self: Violence, Subjectivity, and Early Modern Texts* (JHU Press, 2002), 20.

[57] Marshall, *The Shattering of the Self*, 14.

[58] Marshall, *The Shattering of the Self*, 2.

[59] Marshall, *The Shattering of the Self*, 53.

The Friar he cursed the Knife,
The Friar he cursed the Knife;
He sighs and he groans for the loss of his Stones
He had rather have parted with Life.

In an extraordinary ballad called 'The Gelding of the Devil', a man faces castration because of a deal he has unwisely made with the devil on his way to market. Learning of the danger, his wife tells him

I'd rather thou should break thy Neck-Bone,
Then for to lose any manner of Stone,
For why 'twill be a loathsome thing,
When every Woman shall call thee Gelding.[60]

As we have seen, judicial castration could be associated with the annihilation of the subject as part and parcel of execution. However, these refrains suggest that self-inflicted castration could also affect the annihilation of subjectivity, and specifically masculine subjectivity, even when the person was still alive. Rather than a wholesale 'shattering' of the self, as Marshall suggests of on-stage violence, representations of self-gelding imagined a scenario in which one could excise masculinity, with all its pressures and privileges; a sort of *social* suicide.

This view accords with recent scholarly work on shame in the early modern period. It has been argued that the seventeenth century saw a shift from a shame culture to a guilt culture, or from 'reputation-shame' to 'guilt-shame'.[61] In the former, bad behaviour on the part of an individual is externally recognised and sanctioned, whereas in the latter, an individual recognises and sanctions their own bad behaviour (or as they may understand it, their sin). In reality, however, the experience of shame in early modern England was far more complicated. Certainly, shame often relates to what one has done to other people or what one has been seen *to* do by other people, and in this respect, it is social. However, it is

[60] "The Gelding of the Devil. / OR, / The Prittiest Jest That e're Was Known, / How the Bakers Wife Her Skill Hath Shown, / Then Listen a While, and I the News Will Tell, / Betwixt the Baker and the Devil of Hell". Document. British Library—Roxburghe 2.192–93. 1670, 165.

[61] Hannah Dawson, "Shame in Early Modern Thought: From Sin to Sociability". *History of European Ideas* 45, no. 3 (2019): 377–98. https://doi.org/10.1080/01916599.2018.1534447.

also deeply personal, not least because of its physical effects, including rapid heartbeat, sweaty palms and, most potently, the sensation of blushing. Early modern writers were perennially fascinated by the topic of blushing, which displayed on the human body the immediacy of the humoral connection between emotions and bodily fluids.[62] Moreover, blushing in turn is visible to others as a marker of shame, completing a cycle in which shame recirculates between the individual and others. Shame therefore provides an example of a social affect which reinforces the extent to which one's identity inheres, in phenomenological terms, in 'being-for-others'.[63] Sara Ahmed explains: 'The very physicality of shame—how it works on and through bodies—means that shame also involves the de-forming and re-forming of bodily and social spaces, as bodies "turn away" from the other who witnesses the shame'.[64] Even when one is ashamed of something which nobody has witnessed, the sense of oneself as an object which *can* be objectively viewed by others is integral to the negative affect of shame. In this way, shame rehearses the push and pull between narrative identity and socially constituted identity which Marshall describes, the 'opposite extremes of dissolution and coherence' between which early modern subjectivity was generated.

Hannah Dawson argues that the threat of public shame was central to the development of the seventeenth century's 'mushrooming public sphere' of coffee-houses, salons and societies. By making people conscious of the way their actions were perceived by others, she contends, reputational shame encouraged individuals to regulate their behaviour.[65] However, it is more difficult to see any constructive element to the shame courted by men who self-gelded. When one has completed a permanent bodily alteration, the opportunity to adjust course to become socially acceptable once again is lost. If 'normal' shame tests the boundaries of social acceptability, self-gelding smashes straight through them. Instead, we might think of self-gelding as an exercise in owning the shame one experiences, just as in masochism the victim possesses some discernment over the nature and timing (if not the exact degree) of their pain. The cuckolded, impotent, hen-pecked or sexually incontinent man is ashamed

[62] Jennifer Panek, "*The Nice Valour's* Anatomy of Shame", *English Literary Renaissance* 48, no. 3 (2018): 342–44, https://doi.org/10.1086/699030.

[63] Dan Zahavi, *Shame, Self and Other* (Oxford University Press, 2014), 236.

[64] Sara Ahmed, *Cultural Politics of Emotion* (Edinburgh University Press, 2014), 103.

[65] Dawson, 'Shame in Early Modern Thought', https://doi.org/10.1080/01916599.2018.1534447.

of himself because he fails to live up to the demands of the patriarchal ideal. The gelded man, however, has both the opportunity and the burden of living up to that ideal removed. His liminal status is reflected in the way he is apparently shunned by his community and especially by women, becoming an almost ghostly presence. We are repeatedly told of castrate men that no woman 'will e'er come near him', 'Not a Wench of them all will his presence come near'. Far from creating a new gender identity like the eunuch or castrato, the self-gelded man remains conspicuously outside the reach of any identity category. They find themselves worthy of marginalisation, and then complete the acts which secure their marginalisation by others.

5.4 Conclusion

The kind of self-wounding enacted by men who gelded themselves was in many ways diametrically opposed to other modes of self-injury represented in poems and plays. Where violent lament was serious, representations of self-wounding were jovial, poking fun at the protagonist and those around them. Where dramatic renderings of self-stabbing leaned on heroic antecedents and classical philosophy for their significance, self-gelding was associated with humiliating punishments and shaming rituals. These differences were reflected in the cultural work that was done by representations of self-gelding. It matched failure to live up to the patriarchal ideal with literal emasculation, so that to fail to be the right sort of man was by implication to fail at being a man at all.

While it may seem far removed from the concerns of formal rhetoric, however, self-gelding retained at its heart the same preoccupations about how the self was constituted and experienced in relation to other people. By enacting upon themselves the most extreme version of the punishments which could be imposed for their failure to control their households and themselves, men who self-gelded assumed agency over their own bodies, just as Hieronimo's tongue-biting asserts his agency by appropriating the state's ability to impose bodily destruction. The difference in these cases, however, was that the men in question were not acting in response to an imminent threat by another party. Instead, they reacted to insidious, deeply embedded social judgements which meant that even when one's shortcomings and transgressions stayed secret, they still caused gnawing shame and anxiety. Ironically, living in society meant having one's perception of self socially constituted in a way that was so firmly internalised that it could drive the most extreme and *anti*-social actions.

Whether self-gelding actually worked as an opt-out from the demands of masculinity is debatable. Most of the medical and ballad accounts of these men depict them as abashed, ruing their hasty decision. However, we may wonder if the black humour of these texts is a form of nervous laughter. Unlike Portia or Hieronimo, the self-gelded man is no hero, but he is certainly difficult to place.

CHAPTER 6

Defence: Mortification to Modification

OLLIE: That's great to hear. It seems like many of you have learnt something from each other today. With this in mind, I'd love to hear from you all, what is one thing you are worthy of?
PORTIA: I am worthy of being heard
CARLOS: I am worthy of peoples' time
BEN: I am worthy of being seen for who I really am
TAMBURLAINE: I am worthy of being vulnerable
HELENA: I am worthy of taking up space

> As the Danes over went this lande, their common custome was to have a do with Nonnes where they founde them, for lacke of their owne wyves. Tyll at the last they came to a nondrye (Coldyngham some saye) where all the good systers had cut of their owne noses & their over lappes [top lips] to disfygure them selves, & so to escape the daunger, ye may beleve yt and ye wyll.[1]

The 800-year-old story of a Catholic nun in an isolated part of Scotland would seem of limited interest to early modern English Protestant audiences. One might think that this would be especially the case when the story in question was as horrific as this one—a tale in which innocent women face a choice between being violated by others or grievously mutilating themselves, and in which, furthermore, their sacrifice still ends in death. Yet the story of Saint Ebba proved curiously popular in seventeenth

[1] John Bale, *The first two partes of the actes or vnchast examples of the Englysh votaryes gathered out of their owne legendes and chronycles by Johan Bale* (1551), 52.

A. Skuse, *Hurt Feelings*, Palgrave Studies in Literature, Science and Medicine, https://doi.org/10.1007/978-3-032-04565-2_6

century England, appearing in over thirty texts between John Bale's 1551 *Actes of English Votaryes*, above, and Daniel Defoe's 1724 *Tour thro' the whole island of Great Britain*. In this chapter, I examine why Ebba's story spoke to early modern readers and authors, and what it meant to learn this old tale of self-injury in a new context. I argue that Ebba's self-injury was laudable, but also morally problematic. It was not easily categorised, and the ambiguity was about exactly what *sort* of story this allowed it to relate to numerous different kinds of discourses about pain, mutilation and agency. These included controversial discourses about physical mortification, but also stories in which the modification of one's body was a purposeful act, as in tales of fasting women or unimammarian Amazons. Those narratives opened the door for Ebba's self-injury to be read as a gendered act of self-defence. They also point to the enduring difficulty of categorising self-injury, since the relationship between 'self-harm' and various kinds of body modification remains a subject of much debate.

Ultimately, however, Ebba's story concluded, like many stories of women's self-injury, with her death. In the final section of this chapter, I argue that in early modern culture the self-mutilated woman, even more than the self-wounded man, resisted interpretation. Like the rape victim, she became a figure for whom a social value, and thus a continued existence, could not be imagined. Ebba's mutilation and her death are intertwined, and her nose-cutting may be read as at once agency, abjection and annihilation.

6.1 Identifying Ebba

Confusion surrounds almost every aspect of the story of Saint Ebba. First, her name appears differently in various sources—Ebba, Ebbe, Aebba or Aebbe. Secondly, she is easily confused with another Ebba, also prioress of Coldingham. Ebba the Elder was the daughter of an Anglo-Saxon king, and apparently founded Coldingham monastery in around 642 CE. Like Ebba the Younger, her spirituality was famously entwined with a desire to escape male sexuality. She founded the monastery in part to avoid the attentions of a suitor, Prince Aidan, whom she did not wish to marry, and it was said that due to her prayers, the tide stayed high around Coldingham for three days, preventing him from visiting. It was also alleged that shortly after Ebba the Elder's death in c.683 CE, the priory burned down, in fulfilment of a prophecy which had been made some years earlier. This aspect of the story clearly complicates matters for Ebba the Younger, who is supposed to have occupied the same priory 200 years later, but the discrepancy is not mentioned in early modern texts, which sometimes amalgamate the elder and younger Ebbas into one historical figure.[2]

Regardless of these complications, Ebba's story survived where many other tales of minor saints fell into obscurity. It was recorded in vivid detail in the thirteenth-century *Flores Historiarum*, which was influential on many later histories of the British Isles; in 1603, for instance, the English Catholic Jesuit priest and prolific author Robert Persons cited that text as the source for his observations on Ebba.[3] By the time it began circulating widely in early modern texts, the account was well established. At least thirty-two texts authored between 1600 and 1725 recount the story of Ebba, often as part of general histories of Britain or texts of religious instruction, but less commonly in martyrologies or even in poems. In 1608, for instance, John Wilson's *English Martyrologe* recorded that

> At Coldingham in Norhamshire, in the Marches of Scotland, the passion of S. Ebba Abbesse and Martyr, descended of the bloud royall of the Kings of North*um*berla*nd*, togeather with all her Sisters in the Monastery who in the first Danish Incursions, under the Captayns Hinguar & Hubba, cut of their noses and upper lippes, to deforme themselves, therby to avoyd the barbarous lust of the Pagan persecutours; who seeing them so mangled and defaced, commaunded their Monastery to be fiered, and so they all ended their course of Martyrdome.[4]

[2] See for instance Thomas Fuller, who mixes up the elder and younger Ebbas in this passage: 'Saint Ebba was born in Northumberland, being daughter to Edilfrid the King thereof; When her father was taken prisoner, she got hold of a boat in Humber, and passing along the raging Ocean, she safely landed at a place in Merch in Scotland, which is call'd the Promontory of Saint Ebb unto this day. Becoming Prioress of Coldingham in that Country, to preserve her own and fellow-Nuns chastity from the Pagan Danes, She cut off her own Nose, and perswaded the rest to do the like...' Thomas Fuller, *The History of the Worthies of England* (J.G., W.L. and W.G. for Thomas Williams, 1662), 304.

[3] Matthew Westminster, *The Flowers of History, Especially as Relate to the Affairs of Britain*, transl. by C. D. Yonge (Henry G. Bohn, 1853), 1:410; Robert Persons/Parsons, *An Examen of the Calendar Or Catalogue of Protestan Saincts, Martyrs, and Consellors, Deuised by [J]on Fox, and Prefixed Before His Great Volume, of Acts and Monuments* (1603).

[4] John Wilson, *The English Martyrologe Conteyning a Summary of the Liues of the Glorious and Renowned Saintes of the Three Kingdomes, England, Scotland, and Ireland. Collected and Distributed into Moneths, after the Forme of a Calendar, According to Euery Saintes Festiuity. VVherunto Is Annexed in the End a Catalogue of Those, Who Haue Suffered Death in England for Defence of the Catholicke Cause, since King Henry the 8. His Breach with the Sea Apostolicke, Vnto This Day. By a Catholicke Priest* (1608), 230–31.

The story was still being reproduced to closely the same pattern—though with greater authorial flourish—just over 100 years later, when John Spelman's *Life of Ælfred the Great* told how

> Ebba (afterwards sainted) then Abbess of Coldingham Nunnery in Yorkshire, being more afraid of the Loss of her virginity, than her Life, calling her Nuns into the Chapter-House, made there a most Pathetical Speech to them, setting forth the Lust and cruelty of the Danes, an also exhorting them to follow her Example in avoiding it; which they all unanimously promising to observe, she then took a Razour, and cut off her own Nose and upper Lip, in which she was immediately followed by all the Sisters; which being done, those Cruel tyrants coming thither the next day with their Forces, and seeing so horrible a Spectacle, they not only ran out of the Monastery and left them, but also gave order to their Followers, that they should set the House on Fire, which was punctually put in execution: so it was burnt to Ashes, together with the Abbess and all her Nunns, who thought themselves happy in thus suffering Martyrdom for the Preservation of that which was dearer to them than their Lives.[5]

In the interim between these two works, Ebba's ordeal was described in, among other places, John Bulwer's popular *Anthropometamorphosis* and Helkiah Crooke's *Mikrokosmographia*, as well as in several dictionaries. It was also recalled in Michael Drayton's *Poly-Olbion,* which told how:

> O *Ebba,* whose cleere fame, time never shall out-weare,
> At *Coldingham,* farre hence within that Country plac'd;
> The Abbesse, who to keepe thy vayled Virgins chast,
> Which else thou fearst the *Danes* would ravish, which possest
> This Isle; first of thy selfe and then of all the rest,
> The Nose and upper Lip from your fayre faces kerv'd,
> And from pollution so your hallowed house preserv'd.
> Which when the *Danes* perceiv'd, their hopes so farre deluded,
> Setting the house on fire, their Martyrdome concluded.[6]

[5] John Spelman, *The Life of Ælfred the Great, by Sir John Spelman Kt. From the Original Manuscript in the Bodlejan [Sic] Library: With Considerable Additions, …* (Guil. Lancaster, 1709), 34–35.

[6] Michael Drayton, *The Second Part, or a Continuance of Poly-Olbion from the Eighteenth Song Containing All the Tracts, Rivers, Mountaines, and Forrests: Intermixed with the Most Remarkable Stories, Antiquities, Wonders, Rarities, Pleasures, and Commodities of the East, and Northerne Parts of This Isle, Lying Betwixt the Two Famous Rivers of Thames, and Tweed* (1622), 101.

Whether brief or expansive, each iteration included the same few points: the threat of rape by invading Danes; the fact that the other nuns likewise mutilated their faces (or consented to have them mutilated by Ebba); and the firing of the monastery in revenge. These features remained the same whether the story was recounted by Catholic or Protestant authors. Only one text, James Heath's 1689 *England's Chronicles*, varied significantly from the formula by attesting that the nuns were raped despite their self-mutilation: '[their actions] prevailed not, for the Monsters having first deflowred them, put them to the Sword, and set the House on fire'.[7]

Ebba's story was therefore well-known in this period, and we can reasonably assume that it was familiar to most people with more than a passing interest in Scottish and English history, or the history of the Church. What is less clear, however, is what *kind* of story this was. It was presented as a sort of hagiography, but it was a resolutely unusual one. Since Ebba was sainted pre-Congregation, it is unclear exactly which aspect of her life warranted canonisation, and there is nothing in the account of her death to suggest a miracle has taken place. Nor was this a straightforward story of martyrdom. Though numerous seventeenth texts describe Ebba as 'martyr' (following the example of the *Flores Historiarum*) the issue was complicated; Ebba mutilates herself to preserve her virginity, but it is unclear whether this action contributes to her death, or whether death was inevitable in either case. Moreover, some seventeenth-century authors mingled their admiration for Ebba with ambivalence about her self-sacrifice. In the preamble to Spelman's account, above, he suggests that Ebba's actions indicate 'an Heroick (if not too great a) Love of Chastity'.[8] In so doing, he directly repeats James Tyrell's 1696 account in *The Generall History of England*, indicating that this was an anxiety which circulated from one text to another. Other versions of the story, while not as explicit in their qualms, emphasised the unusual 'zeal' which led Ebba to mutilate herself.[9] On the other hand, as I explore below, several texts suggested that the nuns' beauty was a 'temptation' or 'bait' to the attackers,

[7] James Heath, *Englands Chronicle, or, The Lives & Reigns of the Kings and Queens from the Time of Julius Cæsar to the Present Reign of K. William and Q. Mary* (Printed for Benj. Crayle, N. Bodington, and G. Conyer, 1689), 45.

[8] Spelman, *The Life of Ælfred the Great*, 34.

[9] Daniel Defoe, *A Tour Thro' the Whole Island of Great Britain, Divided into Circuits or Journies. Giving a Particular and Diverting Account of Whatever Is Curious and Worth Observation* (G. Strahan. W. Mears. R. Francklin. S. Chapman. R. Stagg, and J. Graves, 1724), 9.

and thus implied that if they did *not* mutilate themselves the nuns would therefore be complicit in their own rape.[10]

Such ambivalent moments reflect the uncomfortable fact that in much the same way that Hieronimo's death worried at the line between heroism and madness, Ebba's self-wounding and death pushed the boundary between laudable martyrdom and misguided self-destruction. Was the preservation of a virtue a good enough reason to deface God's creation, and endanger one's own life? Various texts approached this question, often with reference to Matthew 5:29:

> And if thy right eye offend thee, pluck it out, and cast it from thee: for it is profitable for thee that one of thy members should perish, and not that thy whole body should be cast into hell.

With increasing interest in the morality and legality of suicide in this period, several influential authors stated that Matthew's words should not be interpreted literally, and that the Christian prohibition on self-killing extended to self-injury, *even* in the pursuit of virtue. In John Sym's text against suicide, for instance, he argued that:

> neither to punish a sin past, nor to prevent a sin to come, may a man destroy or cut off any of his members, whereby he may be lesse able to doe the offices and duties for which God hath given him the same; seeing that both for chastisement, and prevention of sin, God hath appointed other morall meanes, which wee are to use, and therein to depend upon God for the successe; for not in mans forced disability to act sin, but in the renovation of the heart consists true sanctification: that of pulling out the right eye, and of cutting off the right hand, Mat 5.29.30 is meant of morall mortification, whereby these members are made uselesse, and as if they were not to any unlawfull use.[11]

Sym was clear—though one might cut off a limb to preserve life, doing the same to prevent a sin was not allowed. The same sentiment was voiced in contemporary discussion of the church father Origen of Alexandria, who famously castrated himself in order to avoid any accusations of impropriety with his female students. Commenting on the story, John Bulwer

[10] Fuller, *The History of the Worthies of England*, 302.

[11] John Sym, *Lifes Preservative against Self-Killing. Or, An Useful Treatise Concerning Life and Self-Murder* (M. Flesher, for R. Dawlman, and L. Fawne, 1637), 110.

concluded that 'to deprive himself (however sanctimonious his intentions were) of those parts, contrary to the order of nature, was an unlawfull mutilation, and meere treason committed against her'.[12]

Despite this clarity in some quarters, it is less obvious whether mutilation to prevent one being sinned *against*, as in Ebba's case, was similarly forbidden. As I discuss below, the early modern glorification of chastity skewed discourses towards promoting its preservation by any means. Thus when the churchman Clement Barkdale interpreted Matthew 5:29, he noted that though in general it should be taken metaphorically to mean denying one's lusts, 'yet even Maids have with satisfaction chosen, not only to deny themselves the greatest pleasures of the sense, but to sacrifice the seat of them, the body itself, to preserve the satisfaction of being chaste'.[13] Nonetheless, there was at least one authoritative voice warning against suicide to avoid rape. Thomas More's *A Dialogue of Comfort Against Tribulation* (written in c.1534 and published in 1573) drew on St. Augustine's tale of Roman virgins drowning themselves to avoid rape to point out that according to Augustine, 'it is not lawfull for any other maid to follow their sample but rather suffer other to do her any maner [of] violence by force, & commit sinne of his own upon her against her will than willingly & thereby sinfully her selfe become an homicide of her selfe'.[14] Once again, More's advice did not specifically cover Ebba's actions, but it accorded with a trend in contemporary moralistic texts of emphasising the importance of submission to God's will and reiterating the impossibility of ensuring one's own salvation through deeds.

[12] John Bulwer, *Anthropometamorphosis: = Man Transform'd: Or, the Artificiall Changling* (Printed by William Hunt, 1653), 359.

[13] Clement Barksdale, *Three Ministers Communicating Their Collections and Notions. Touching Several Texts of Scripture … Wherein the Law and Gospel … in Short, the Substance of Christianity Is Set Forth …* (Printed for Samuel Keble, 1675), 127. See also Alanna Skuse, "'One Stroak of His Razour': Tales of Self Gelding in Early Modern England". *Social History of Medicine* 33, no. 2 (2020): 377–393. https://doi.org/10.1093/shm/hky100.

[14] Thomas More, *A Dialoge of Comfort against Tribulacion, Made by Syr Thomas More Knyght, and Set Foorth by the Name of an Hu[n]Garie[n], Not before This Time Imprinted* (1553), fols. K1v–K2r.

6.2 'Is this Mortification, to murther our selves?'[15]

The problems of categorisation which attended Ebba's story (good or bad, brave or foolish) reflected long-running uneasiness about stories of religiously motivated self-injury. Stephen Greenblatt observes that in the Christian corpus there was undeniably '[a] link between mutilation, as a universal emblem of corporeal vulnerability and abjection, and holiness'.[16] That link found expression partly through the stories of martyrdom into which Ebba's tale was imperfectly fitted. Such stories were undoubtedly popular throughout the medieval and early modern period; accounts such as Ebba's demonstrate how saintly tales continued to circulate even when the saint themselves no longer had much place in religious teaching. Moreover, the bloody conflicts of the Reformation breathed new life into the concept of martyrdom, largely through John Foxe's popular—and lengthy—*Actes and Monuments.*[17]

However, both the moral problematics and the uniquely communal nature of the self-wounding practised by Ebba and her followers must also have evoked another kind of religious pain, from which medieval and early modern readers were rather less removed; that is, the pain of physical mortification. Mortification had long been important in the Christian faith. Esther Cohen describes how in the later Middle Ages, both men and women 'practiced the self-infliction of pain and welcomed divinely inflicted sufferings', while 'during the thirteenth century, collective flagellant processions began appearing in Europe, initiating a movement of communal self-mortification'.[18] Cohen notes that this practice was meant to imitate

[15] John Gee, *The Foot out of the Snare: With a Detection of Sundry Late Practices and Impostures of the Priests and Jesuits in England* (H. Lownes for Robert Milbourn, 1624), 81.

[16] Stephen Greenblatt, "Mutilation and Meaning". In *The Body in Parts: Fantasies of Corporeality in Early Modern Europe*, ed. David A. Hillman and Carla Mazzio (Routledge, 1997), 223.

[17] See John N. King, "Introduction," in *Foxe's "Book of Martyrs" and Early Modern Print Culture,* ed. by John King (Cambridge University Press, 2006), 1–20, https://doi.org/10.1017/CBO9780511483646.002; Susannah Brietz Monta, *Martyrdom and Literature in Early Modern England* (Cambridge University Press, 2005); Anne Ruth Throckmorton, "The Persistence of Saints in Early Modern England" (Ph.D., United States -- Virginia, University of Virginia, 2011).

[18] Esther Cohen, *The Modulated Scream: Pain in Late Medieval Culture* (University of Chicago Press, 2009), 27.

the suffering of Christ, and explicitly *not* to mar the divine creation through self-mutilation. For almost as long as it had been practised, however, mortification had been problematic. Even in the medieval heyday of physical self-mortification, some commentators were apparently aware that the bounds within which such practices ought to operate were often hard to define. For example, the thirteenth-century *Ancrenne Wisse*, a book of monastic rule, warned that a nun

> should not wear a belt of any kind next to her skin except with her confessor's permission, or wear anything made of iron or hair or hedgehog skins, or beat herself with them, or with a scourge weighted with lead, with holly or with thorns, or draw blood, without her confessor's permission. She should not sting herself anywhere with nettles, or scourge the front of her body, *or mutilate herself with cuts*, or take excessively severe disciplines at any one time to subdue temptations.[19]

This excerpt shows the caution with which mortificatory practices were viewed even within an obviously devotional context. Tight control over the manner, place and time of such acts was crucial if one was to maintain a distance between Christian mortification and the self-sacrificial practices of 'heathen' cultures such as Muslims.[20] As Claire Marshall has described, however, religious self-injury always threatened to slip the control of the authorities.[21] Indeed, in its most visible form—the flagellant procession—it was often explicitly anticlerical, and such activities were banned by Pope Clement VI in 1349.

With the advent of the Reformation, self-mortification was easily situated by Protestant authors as a practice for self-indulgent Catholics. Greenblatt describes how

> in 1532 Luther's great associate Philipp Melanchton, commenting on Romans 12:1—"present your bodies a living sacrifice, holy, and acceptable to God"—urges his readers to "think that we always stand bound and garlanded at the altar, reaching our necks out to enemies who strike us." But, Melanchton quickly adds, though Saint Paul commands us to be sacrifices

[19] Bella Millett and Jocelyn Wogan-Browne, *Medieval English Prose for Women: Selections from the Katherine Group and Ancrene Wisse* (Oxford University Press, 1992), 137. My emphasis.

[20] Greenblatt, "Mutilation and Meaning", 229.

[21] Marshall, "The Politics of Self-Mutilation".

> for Christ's sake, "it is not because we injure or destroy ourselves but because others injure us". For the Reformer, Catholic self-mortification is one of a range of practices that are neither holy nor acceptable to God.[22]

This antipathy to practices such as flagellation or the wearing of hair shirts only hardened over time. By the seventeenth century, English authors routinely emphasised that mortification was only to be celebrated when it was entirely spiritual in character: meditating on one's own sinfulness, considering the sufferings of Christ and so forth. Writing on flagellants in 1637, William Watts complained that: 'Such barbarismes, surely, can no more appease God, then Baals priests could, by cutting themselves with rasors and with lances', pointedly adding 'If it be an act of devotion, why then is it acted in a Procession?'[23] Thomas Wolfall likewise identified 'Papist' flagellants as having 'farre more blind devotion than true discretion'.[24] Physical mortifications, he explained, 'are but like so many outward medicines, wich will be unable to cure an inward disease, this is but, as it were, to apply a plaster to the head, when your griefe lyeth at the heart'.[25]

As the century wore on, the conviction among authors that bodily mortification was a Popish practice grew stronger, such that in 1668, John Owen's 'Of the mortification of sin in believers' asserted that

> The greatest part of Popish Religion, of that which looks most like Religion in their Profession, consists in mistaken ways and meanes of mortification. This is the pretence of their rough garments, whereby they deceive. Their Vows, Orders, Fastings, Penances, are all built on this ground, they are all for the mortifying of Sin … the Friers of the Romish church, who are said to torment men so that they should seek death and not find it, think, that they did it by their stinging Sermons, whereby they convinced them of sin, but being not able to discover the Remedy for the healing and Mortifying of it, they kept them in such perpetual anguish and Terror … that they desired to dye.[26]

[22] Greenblatt, "Mutilation and Meaning", 230.

[23] William Watts, *Mortification Apostolicall* (John Legat for John Cowper, 1637), 16.

[24] Thomas Wolfall, *The Doctrine and Practice of Mortification* (T.C. for John Sweeting, 1641), 105–06.

[25] Wolfall, *The Doctrine and Practice,* 105–06.

[26] John Owen, *Of the Mortification of Sin in Believers*, 3rd edition. (Nathanael Ponder, 1668), 27.

Owen's text highlighted numerous anxieties about mortification and its links to extravagant 'stinging' sermons. His emphasis on the 'look' and 'pretence' at which 'vows, orders, fasting, [and] penances' are aimed repeats the suspicion articulated by Watts when he asked why flagellation should be *acted in a procession*. Both view physical mortification as ostentatious; while this was a common accusation levelled at Catholicism in general, it also echoed anxieties present in secular representations of self-wounding, wherein acts of self-injury might easily be construed as overly performative and thus inauthentic. Owen's concern about the 'perpetual anguish' which drove lay Catholics to desire extremes of mortification which they could not achieve also suggests a concern with the deliberate stirring up of emotions similar to that which troubled some critics of violent lament. Taken together, these misgivings show how physical mortification occupied an uncertain position in relation to those discourses promoting stoicism and temperance which flourished in the early seventeenth century. On the one hand, mortificatory practices certainly demonstrated self-mastery over the body. On the other, there was a feeling that mortification aggrandised the mind as much as it subdued the flesh. Mortification nourished unhealthy levels of despair and self-loathing, but it also enabled personal pride in the fortitude with which one could express those feelings.

Despite mortification's problematic elements, it never entirely died out. Robert Schneider has written about how in northern Italy and Provence, medieval flagellation was later 'channeled into confraternities of *disciplinati*', and in this guise 'remained an important aspect of processional display throughout the Renaissance and Counter-Reformation'. Moreover, 'In the sixteenth century, this organized form of penitential ceremony swept much of France, where it flourished amidst the turmoil of the religious wars, becoming a highly effective means of stimulating Catholic militancy and fervor'.[27] Even in seventeenth-century England, communities of Catholics occasionally persisted in public mortificatory practices, as when, in 1624, a procession walked to Tyburn naked from their waists upward, scourging themselves with whips. The parade was recorded by John Gee, who decried the practices in part because of their potential to

[27] Robert A. Schneider, "Mortification on Parade: Penitential Processions in Sixteenth- and Seventeenth-Century France". *Renaissance and Reformation* 22, no. 1 (1986): 123–46, (123).

spill over into lasting mutilation or even suicide. Addressing the flagellants, he chided that

> you made one whip himself so long, till he swouned, and was thought to be past hope of recovery, so that hot water was instantly fetched to revive him. At Bruxels (as a Priest told mee, saying hee saw it, and boasting of the meritorious work) a woman, about a yeere since, so cruelly scourged her selfe, that shee died of it.[28]

For Gee, as for Owen, this was not so much penitence as 'unnaturall hating and disabling [of] our corporal faculties'.[29] Yet it was also a spectacle which made good material for popular texts. The idea of mutilating or mortifying the flesh remained imaginatively compelling long after it became theologically defunct.

6.3 Religion, Power and Starvation

Given the long association between self-inflicted pain and religious belief, one might expect that many reports of self-injury in early modern England would cite faith as a factor. Yet in fact, very few narratives of self-wounding mention religion at all. Moreover, the handful that do mention faith strongly resist any meaningful connection between self-injury and piety. In 1679, for example, the surgeon James Yonge was called to an incident in which a young man had cut off his testicles. The patient told Yonge that he had acted 'on a religious account, to cure salacious heats'.[30] However, Yonge found this an unconvincing explanation. Though the young man might be 'acting *Origen Secundus*' in his methods, said Yonge, he thought it likely that he was motivated by 'some disappointment in love' rather than any real religious fervour.[31] The connection between religion and self-injury remained present in certain accounts of people who hurt themselves during episodes of madness, but such episodes were represented as melancholy delusions, in which it was more likely that the patient

[28] John Gee, *The Foot out of the Snare: With a Detection of Sundry Late Practices and Impostures of the Priests and Jesuits in England* (H. Lownes for Robert Milbourn, 1624), 81–82.

[29] Gee, *The Foot out of the Snare,* 82.

[30] James Yonge, *Currus Triumphalis, è Terebinthô. Or An Account of the Many Admirable Vertues of Oleum Terebinthinæ* (Printed for J. Martyn, 1679), 76–78.

[31] Yonge, *Currus Triumphalis.*

perceived themselves to be prompted by the devil than by God. Probably the most striking example of this was Hannah Allen's narrative of her long struggle with melancholy, in which she repeatedly poisoned, starved and injured herself, motivated by belief in her own wickedness.[32]

Religious self-sacrifice or self-mortification did not therefore map directly onto secular representations of self-wounding; it was not the case that authors scratched flagellant processions out of their works and replaced them with Stoic self-injurers or violent lamenters. However, this is not to say that the two phenomena were entirely divorced from one another. Instead, some of the broad themes which had attended representations of martyrdom and mortification now emerged in stories where religion was not a prominent aspect. A useful precedent here is that of food refusal, a phenomenon which has attracted far more scholarly attention than self-stabbing, auto-amputation, or other kinds of violent self-injury. Even when it was rooted in religious experience, this practice retained the potential for anti-authoritarianism. In Caroline Walker Bynum's foundational study of food refusal among medieval women, she finds that fasting was often both a genuine act of religious devotion *and* a way in which women could assert themselves. Religious women, and nuns in particular, practised various kinds of asceticism, from going barefoot in winter, to rubbing themselves with pus from plague victims, to rolling in broken glass.[33] However, since women were commonly in charge of food preparation, it made sense that food would be the medium through which both religious and social desires were most commonly expressed, and thus there are many accounts of lengthy fasting by both lay and cloistered women. Bynum explains:

> First, women's food behavior—fasting and feeding—was an effective way of manipulating the environment in a world in which food was woman's primary resource. Second, women's radical asceticism was less an internalizing of the church's negative views of flesh and female than a rebellion against the moderation of the high medieval church, which was moving toward a more positive sense of the body. Third, food asceticism, food distribution,

[32] Hannah Allen, *A Narrative of God's Gracious Dealings with That Choice Christian Mrs. Hannah Allen (Afterwards Married to Mr. Hatt,) Reciting the Great Advantages the Devil Made of Her Deep Melancholy, and the Triumphant Victories, Rich and Sovereign Graces, God Gave Her over All His Stratagems and Devices* (Printed by John Wallis, 1683).

[33] Caroline Walker Bynum, *Holy Feast and Holy Fast: The Religious Significance of Food to Medieval Women* (University of California Press, 1988), 210.

> and eucharistic devotion did not, to medieval people, mean self-torture; rather, they were ways of fusing with a Christ whose suffering saves the world.[34]

This dual social and spiritual practice is visible in the much-studied case of Margery Kempe, the early fifteenth-century mystic who practised fasting among other forms of asceticism and devotion. Kempe believed her fasting to be supported by Jesus and had visions to that effect.[35] However, her practice was also powerfully subversive. Kempe's vision of having free access to the Eucharist (usually tightly controlled by the Church) was 'co-opted' into contemporary debates about the doctrine of transubstantiation.[36]

Leanore Leiblein and Christopher Frey have shown that in the early modern period, interpretations of food refusal shifted, albeit slowly.[37] Looking in particular at literary representations of women's fasting, they argue that

> The medieval belief in starvation as a means to higher spiritual and moral calling remained current in [Thomas] Heywood's time [i.e., early 1600s]. Bartholomew Batty, for example, bases his argument in favour of fasting on the belief that 'the Soule and minde is heavenly: but the Body wee have in common with Beastes.'... By the end of the seventeenth century, however, fasting had become a medical and scientific rather than a purely spiritual phenomenon ... For late seventeenth-century observers of female starvation, the woman in question is no longer of interest for speculation about the relationship between her body and soul and mind; instead, her body is an object of interest in and for itself.[38]

Religious readings of starvation remained in play for a long time. Examining the famed 'fasting maiden' Martha Taylor in 1669, for instance, commentators went back and forth on the likely cause of her ability to live without food for a year. Thomas Robin declared that 'I do believe she is a chosen vessel of the Lord's'.[39] The author of *Mirabili Pecci* pondered both

[34] Bynum, *Holy Feast*, 218.

[35] Marshall, "The Politics of Self-Mutilation", 17.

[36] Marshall, "The Politics of Self-Mutilation", 18.

[37] Christopher Frey and Leanore Lieblein, "'My Breasts Sear'd': The Self-Starved Female Body and A Woman Killed with Kindness," *Early Theatre* 7, no. 1 (2004): 45–67 (47).

[38] Frey and Lieblein, "My Breasts Sear'd", 46.

[39] Thomas Robins, *The Wonder of the World* (Tho[mas] Passinger, 1669), 15.

Martha's devotion, which they found to be sincere, and the possibility of her being sustained by humid airs.[40] Thomas Hobbes asked questions and concluded that 'I cannot ... deliver any Judgment in the case', while John Reynolds was firmly of the opinion that this was a puzzle for scientists, not clergymen.[41] Despite debate over individual instances, however, the phenomenon came to be broadly viewed as 'not ... saintly, but sickly'. Yet, this shift in interpretive emphasis did not diminish the agential potential of self-starvation, only redirected it to more secular aims. Considering the raft of commentary provoked by the death from starvation of Margaret Radcliffe, a lady-in-waiting to Elizabeth I, Nancy Guitterez notes that

> Stories and plays about female food refusal overtly demonstrate a social order that is able to control the potential unruly nature of a woman's body, *at the same time as* each of the stories suggest that social order is unable to control that body. These coincident oppositional meanings allow us to see the distinctive stress that an expression—or perceived expression—of human agency places on institutional authority in early modern culture.[42]

Leiblein and Frey note other examples (both real and fictional) of food refusal as agential. In Heywood's *A Woman Killed with Kindness,* for instance, they view the adulterous wife Anne's food refusal, from which she eventually dies, as not only penitential but emancipatory: 'By voluntarily self-starving, Anne violates the body perceived as property, and resists and successfully eludes the male-determined symbolic order'.[43]

[40] H.A., *Mirabile Pecci, or, The Non-Such Wonder of the Peak in Darby-shire* (T. Parkhurst and G. Calvert, 1669), 62.

[41] Thomas Hobbes, "Letter 183 20 [/ 30] October 1668 Thomas Hobbes to John Brooke, from Chatsworth". In *The Clarendon Edition of the Works of Thomas Hobbes, Vol. 7: The Correspondence, Vol 2. 1660–1679*, ed. Noel Malcolm (Oxford University Press, 1994); John Reynolds, *A Discourse Upon Prodigious Abstinence: Occasioned by the Twelve Months Fasting of Martha Taylor, The Famed Derbyshire Damosell: Proving That Without Any Miracle, the Texture of Humane Bodes my be so altered, that Lafe may be Long Continued without the Supplies of Meat & Drink* (Printed by R.W. for Nevill Simmons and Dorman Newman, 1669).

[42] Nancy A. Gutierrez, '*Shall She Famish Then?': Female Food Refusal in Early Modern England* (Routledge, 2016). 34. On Radcliffe, Gutierrez notes that she starved herself apparently from grief over the death of the brother, and an autopsy suggested that she had died from a broken heart rather than poisoning (stripes seen on the outside of her heart were taken to be broken tendons or 'heart-strings'). This, along with the other-directedness of her emotional distress, allowed her death to be read as noble rather than sinful, and she was the subject of an epitaph by Ben Jonson.

[43] Frey and Lieblein, '"My Breasts Sear'd"', 62.

Likewise, they note the case of Elizabeth Cary, who was left destitute after separating from her husband in 1627 and refused to eat while awaiting the outcome of her petition to Charles I.[44] These women no longer positioned fasting explicitly as a means to draw closer to God. However, like medieval fasting maidens, women who refused food in the early modern period used their bodies as 'a site of political apprehension and cultural change'.[45] Rather than exerting power through interpersonal violence, they took the feminine virtues of chastity, silence and obedience and transformed them into radical displays of self-mastery in which the subjection of the body to the will was writ large.[46]

Ebba's story was thus read, and circulated, in a culture in which understandings of female asceticism were shifting in ways which allowed ever more forensic examination of female physicality, and yet continued to leave room for women to manipulate the social meanings of their bodies. The parallels between such figures and Ebba were not exact. After all, fasting women differed from women who injured themselves because the latter actively harmed their bodies, while the former passively allowed them to decline. Moreover, fasting women's experiences were highly individualised; while they wrought social effects, they did so through the practical means of sealing off the body from the world. Stories about Ebba, as I shall discuss, stress the way in which her mutilation provided an example to her sisters. Nonetheless, fasting—along with self-mortification—provided a well-established context linking bodily self-sacrifice with self-assertion. The continued debate around these practices, and thus about women's proper use of their own bodies, helped to ensure that Ebba's story remained relevant long after her untimely death.

[44] Frey and Lieblein, '"My Breasts Sear'd"', 50.

[45] Gutierrez, *Shall She Famish Then?*, 20.

[46] Both contemporary sources and later critics have noted that most self-starvation in this period was undertaken by young women. Reynolds, for instance, wrote that 'most of these Damosels fall to this abstinence between the age of fourteen and twenty years' (Reynolds, *A Discourse Upon Prodigious Abstinence*, 21). However, there were male examples of fasting, most notably that of Roger Crabb, the 'English Hermite' who espoused a vegan diet and lived for periods of time on grass alone. Crabb used his fame to amplify his political and religious beliefs, and has been described as an 'opposition hunger artist' by Rick Bowers ("Roger Crab: Opposition Hunger Artist in 1650s England". *The Seventeenth Century* 18, no. 1 (2003): 93–112. https://doi.org/10.1080/0268117X.2003.10555521.)

6.4 'Religious Amazons'

Texts which centred religiously motivated flagellation, fasting and asceticism often asked variations of the same question: When were these practices self-sacrificing, and when were they self-serving, or even self-aggrandising? In the seventeenth century, this question was joined by another: Where was the line between harming the body and simply changing it? This is a query which has often been asked in modern writings on self-harm, as many psychologists have identified that the differentiation of 'body-marking' practices into 'healthy' and 'pathological' sub-categories is a normative, and inherently political, process (the DSM-5 specifies that self-injury disorder only includes actions that are not 'socially sanctioned').[47] Almost as many have sought to show either that body modification and self-harm arise from the same impulse, or that body modification may lessen the risks associated with self-harm by providing a comparatively safe outlet for the need to alter one's bodymind, and I discuss this view of self-injury as 'useful' in Chap. 8.[48] Conversely, some sociologists of self-injury have sometimes proved keen to distance that activity from body modification by arguing that whereas modifications are clearly communicative, self-injury is more closely associated with a need to feel pain.[49]

As we have seen thus far, wounding oneself in the early modern period was widely understood as at least partly a communicative practice, and hence its relationship to bodily modification was arguably even more vexed than today. Moreover, it was a relationship brought into focus by the increasing capacity of surgery to transform the body. As I have explored

[47] Kay Inckle, "Telling Tales? Using Ethnographic Fictions to Speak Embodied 'Truth'". *Qualitative Research* 10, no. 1 (2010): 27–47. https://doi.org/10.1177/1468794109348681, 27–28.

[48] Mara Iannaccone and Stefania Cella, Paolo Cotrufo, Federico Manzi, Sabato Antonio Manzi, and Luigia Visconti, "My Body and Me: Self-Injurious Behaviors and Body Modifications in Eating Disorders-Preliminary Results," *Eating Disorders* 21, no. 2 (2013): 130–39, https://doi.org/10.1080/10640266.2013.761087; Alexandra L. Vizgaitis, "Pierced Identities: An Investigation of the Relationships between Body Modification Practices, Borderline Personality Pathology, and Identity Diffusion" (ProQuest Dissertations Publishing, 2015); Aglaja Stirn and Andreas Hinz, "Tattoos, Body Piercings, and Self-Injury: Is There a Connection? Investigations on a Core Group of Participants Practicing Body Modification," *Psychotherapy Research* 18, no. 3 (2008): 326–33, https://doi.org/10.1080/10503300701506938.

[49] Lindsey M. Ibañez and Jim D. Taylor. "Sociological Approaches to Self-Injury". *Sociology Compass* 9, no. 12 (2015): 1005–14, (1007).

elsewhere, procedures such as the creation of castrato singers through orchidectomy raised the possibility that acts which were clearly harms in one context could be experienced as beneficial in another.[50] The greatest source for such examples, however, was in the ever-expanding genre of ethnography, which told tales of distant peoples for whom bodily 'mutilations' were in fact expressions of group identity and aids to beauty or function. Plentiful examples of such practices were provided in John Bulwer's immensely popular *Anthropometamorphosis*. Printed in 1650 and again in an expanded edition in 1653, this text depicted natural 'monsters' (men with their faces in their stomachs, covered in hair like beasts, improbably tall or diminutive) and monsters of behaviour (such as cannibals). However, it also depicted scores of people who changed their bodies through self-inflicted injuries such as scarification, tattooing or the elongation of the lips or earlobes. Of these, perhaps the most fascinating—with a surprising relevance to the Ebba story—were the Amazons.

Amazon women had long been the topic of texts of ethnography and history including Thomas Heywood's 1624 *Gynaikeion,* Walter Raleigh's 1596 *Discovery of the Large, Rich, and Beautiful Empire of Guiana,* and John Mandeville's *Voyages and Travels* (printed in 1582, with numerous translations and editions from the fourteenth century onward).[51] In the seventeenth century, Amazons also appeared in various plays—most famously, Shakespeare's *A Midsummer Night's Dream* (1605), but also John Weston's *The Amazon Queene* (1667), John Fletcher's *The Sea Voyage* (1622) and Jasper Maynes's *The Amorous Warre* (1648). Nonetheless, details about this tribe remained hazy. Some accounts placed them in Guiana, others in the Congo. Some, such as Heywood, believed them to have originally been Scythian women who migrated while their husbands were away at war and formed their own society, while others believed that the Amazons were a separate tribe who warred with the

[50] Skuse, *Surgery and Selfhood.*

[51] Thomas Heywood, *Gynaikeion: Or, Nine Bookes of Various History. Concerninge Women Inscribed by Ye Names of Ye Nine Muses* (1624), 215–23; Walter Raleigh, *Discovery of the Large, Rich, and Beautiful Empire of Guiana, by Sir W. Ralegh: With a Relation of the Great and Golden City of Manoa (Which the Spaniards Call El Dorado), Etc. (1596)*, Hakluyt Society, First Series (Ashgate Publishing Ltd, 1848), 27–29; John Mandeville Sir, *The Voyages and Travailes of Sir John Maundevile Knight Wherein Is Treated of the Way towards Hierusalem, and of the Mervailes of Inde, with Other Lands and Countries* (Printed by Thomas Este, 1582), fol. Lii-Liii.

Scythians or bore children by them.[52] The effect of this uncertainty was to allow speculation to flourish, especially about the most distinctive physical feature of the Amazons—their removal of one breast. This was the 'metamorphosis' on which Bulwer's text focussed.

Texts about Amazons usually included references to monomasty. While Raleigh wrote that 'that they cut of[f] the right dug of the brest I do not finde to be true', most other accounts disagreed.[53] Mandeville believed that Amazon girls of noble birth had the left breast removed by cautery to allow them to better bear a shield, while ordinary girls had the right breast removed to befit them for a bow and arrow.[54] Heywood asserted that 'the right brest they burne off, that with the more facilitie they may draw a Bowe, thrill a Dart, or charge a Launce'.[55] In *Anthropometamorphosis*, Bulwer elaborated on those tales and approached Amazonian bodies with a forensic interest:

> The Ancient *Amazones*, of whom we read so oft in learned Authors, were wont to seare off their right breasts, which was then the archers fashion. *Porta* saies, the Amazons sear off their right paps, that more nourishment going into the hand next to it, might increase the strength of that which was but weake by nature. Others say, that the Amazons much helping themselves in the wars with bows and arrows, and finding that in this and other exercises of armes their dugs or breasts were a very great hinderance to them, they used to burne off the right pap, both of themselves and their daughters, and thereupon they were called Amazons, which signifieth in the Greeke tongue, No Breasts.
>
> The chiefe of the guard of the King of Congo are lefthanded Amazons, who seare off their left paps with a hot iron, because it should be no hinderance to them in their shooting. Pigasetta in his reports of the Kingdom of Congo, makes the like mention of these Amazons, who serve the Emperour Monomotapa. Grimston makes mention of these women warriours serving this Emperour, who after the manner of the ancient Scythish [Scythian] or Asiatique Amazons, have their breasts cut off.[56]

[52] Heywood, *Gynaikeion*, 220; Gaultier de Coste La Calprenède, *Hymen's Prœludia, or Loves Master-Peice Being That so Much Admired Romance, Intituled Cleopatra: In Twelve Parts* (W.R. and J.R, 1674), 159.

[53] Raleigh, *Discovery*, 28–29.

[54] Mandeville, *The Voyages*, fol. L2v.

[55] Heywood, *Gynaikeion*, 223.

[56] John Bulwer, *Anthropometamorphosis: = Man Transform'd: Or, the Artificiall Changling* (Printed by William Hunt, 1653), 321–23.

Bulwer's report sought to pin down the Amazons in a way which no previous account had managed. His history reads like a taxonomy, and dwells on the way in which the Amazons' bodily difference is adaptive to their circumstances, with right or left breast removed to fit their needs. It is accompanied by illustrations of the modified women naked from the waist up; images which, like many in Bulwer's text, served both to educate and to titillate.

Immediately following this passage, however, Bulwer changed tack and approached monomasty as a moral issue—a mutilation rather than a modification. Early in the *Anthropometamorphosis* he had argued that man was made in the image of God, and that alterations to His creation were therefore necessarily diminutions. This applied to women too, since 'The breasts by nature are two', and removing one for an 'unnaturall conveniency' had a knock-on effect on the rest of the body:

> for they were ordained to be a kind of covering and defence for the heart, and that themselves having received heat and cherishment from the heart, might again return unto its warmth, such as we get by garments we buckle about us … Another penalty of their crime against the offended Majesty of Nature they must needs incur, unlesse with their breasts they put off the very Nature of woman, since another use of the Paps, according to *Hippocrates*, was, to receive excrementitious moisture: For if (saith *Hippocrates*) any disease, or other event, take away a Woman's paps, her voice becomes shriller, she proves a great spitter, and is much troubled with the pain in her head.[57]

Bulwer's complaint emphasised the 'leakiness' which Gail Kern Paster has identified as a feature in male representations of the female body during this period, and which was a general facet of the way in which women were perceived as slaves to their 'excrementious' and volatile bodies.[58] However, this was not only a medical concern. Bulwer's belief that the Amazons' bodily modification would offend nature, and that this constituted an immoral *mutilation*—even a 'crime'—drew on the same concerns about female bodily agency which often emerged from stories of women's fasting. As Leiblein and Frey point out, both the Amazon and the self-starved woman remove their capacity to feed offspring, a connection which became explicit in certain explanations of the word 'Amazon':

[57] Bulwer, *Anthropometamorphosis,* 324–25.

[58] Paster, *The Body Embarrassed*, 64–112.

> The connection between breast, nourishment, and generation is made explicit by Thevet's observation, 'Others take the *Etymologie* of this vowell *A*. privative, and of *Maza*, that signifieth bread, for that they live not with bread … as those that have been nourished w[ith]out womans milke'. Amazons demonstrate symbolically a willingness to starve themselves, even to extinction, since their removal of breasts and men puts them on a precarious progenitive threshold.[59]

The Amazon, like the fasting woman, undermines their role as nourisher of others at the same time as they refuse nourishment themselves. Unlike the fasting woman, however, the Amazon acts as part of a group, and it was the link between that group's gynocentrism and their bodily difference which often seemed most concerning to commentators. According to Mandeville, the Amazons would famously 'suffer no men to live among them, nor to have rule over them', instead periodically inviting men to copulate with them.[60] This notion was expanded by Raleigh to include the idea that when the Amazons took male prisoners of war, they would use them for sex before eventually killing them.[61] Moreover, they would either abandon their male infants entirely, or give them to other tribes to raise, a repudiation of maternal instinct symbolically linked to their decision to limit their milk-producing abilities. Like the stripes of medieval flagellants, the monomasty of the Amazonian woman was thus no longer only a means to a practical end, but a way of asserting belonging within a community of individuals, whose belief in their collective identity was amplified by the injuries they inflicted upon themselves. In common with many of the other kinds of self-wounding described in this book, for Amazons (at least as they were represented in early modern texts) the practical utility of mastectomy was matched, and often superseded, by its symbolic power.

* * *

While the sex lives of Amazons may seem little related to the ascetic practices of medieval nuns, the notion of *shared* bodily difference as an expression of community and agency was one which repeatedly emerged in seventeenth-century retellings of the Ebba story. Varied accounts

[59] Frey and Lieblein, '"My Breasts Sear'd"', 52.
[60] Mandeville *The Voyages,* fol. L3r.
[61] Raleigh, *Discovery,* 27–29.

concurred that every nun in Ebba's sisterhood underwent the same mutilation. Some authors, such as Helkiah Crooke, thought that Ebba had wielded the knife on her fellow nuns, an act of leadership and brutality which recollected the Amazons' treatment of their daughters.[62] More commonly, however, authors suggested that Ebba had simply used a rousing speech to persuade her sisters to mutilate themselves. This idea—along with others in the Ebba legend—owed much to *Flores Historiarum*, which claimed to reproduce Ebba's speech directly:

> [Ebba] convened all the sisterhood to a chapter, and burst forth in the following words, saying, 'There have lately come into our district most wicked pagans, men ignorant of every kind of humanity, who traverse every part of this region, and spare neither the sex of females nor the age of infants. They trample upon churches and ecclesiastical persons, they defile women bound by the holiest vows, and crush and destroy whatever they meet with. Therefore, if you are willing to adopt my advice, I entertain certain hopes, founded on the divine mercy, that we shall be able to escape the rage of the barbarians, and to preserve the modesty of our perpetual maidenhood.' And when the entire assembly of the virgins had given her their pledge and most solemn promise that they were desirous to obey her maternal injunction in all things, then that abbess of admirable courage, openly giving to all the sisters an example of chastity which should be profitable not only to those nuns, but which should be worthy of being followed by all succeeding virgins, and by all who should at any time exist, took a razor, and cut off her nose, and her upper lip close to the very teeth, and so made herself a shocking sight to all the bystanders. And as the whole assembly which beheld this memorable deed admired it, they all inflicted similar treatment on themselves, and followed the example of their mother.[63]

In this account, the actual content of Ebba's speech was less remarkable than the persuasive effect it had on those who witnessed it. Like Shakespeare's Portia, Ebba moved deftly between speech and action for maximum rhetorical impact, and the emphasis here is less on the effect that her mutilation had on the Danes, and more on the 'shocking', 'memorable' impression it made on the other nuns. Moreover, Ebba's speech emphasises the gynocentric status of her community, in which a leader is also a 'mother'. While Ebba is chaste, her example is endlessly

[62] Helkiah Crooke, *Mikrokosmographia a Description of the Body of Man. Together With the Controversies Thereto Belonging* (William Jaggard, 1615), 650.

[63] Westminster, *The Flowers of History*, 1:410.

reproductive, inspiring not only the nuns present, but 'all succeeding virgins' and finally 'all who should at any time exist'.

Though later accounts of the Ebba story were usually (but not always) less florid, they repeatedly returned to the extraordinary leadership Ebba exerted over her community. Six seventeenth-century texts referred to Ebba as having 'perswaded' her nuns, while others said that she 'exhorted' or 'advised' them to follow her example.[64] In 1623, John Abbot's five-book religious poem *Jesus Prefigured* provided an expansive vision of Ebba's dilemma and sacrifice. The way in which Abbot chose to embellish the original tale was intriguing for its consideration of self-mutilation as an act which could be both assertive (of one's own rights over the body) and submissive (to the will of God). Having heard of impending danger and fled to the church, Ebba apparently has little time to address her nuns before the Danes break down the door, but still manages two verses of stirring rhetoric:

> (daughters) quoth shee
> To free your selfe from Danes lust learne of mee;
> What law forbids to use a murdring hand,
> To keepe vowd'e faith, the same law doth commaund
> For beauties sake Pagans have us in chase,

[64] Richard Smith, *The Prudentiall Ballance of Religion Wherin the Catholike and Protestant Religion Are Weighed Together with the Weights of Prudence, and Right Reason* (François Bellet, 1609), 315; John Stow, *The Abridgement of the English Chronicle, First Collected by M. John Stow, and after Him Augmented with Very Many Memorable Antiquities, and Continued with Matters Forreine and Domesticall, Unto the Beginning of the Yeare, 1618. by E.H. Gentleman.* (Edward Allde and Nicholas Okes for the Company of Stationers, 1618), 34; William Slatyer, *The History of Great Britanie from the First Peopling of This Island to This Present Raigne of or [Sic] Happy and Peacefull Monarke K: Iames* (Printed by W: Stansby, for Rich: Meighen, 1621), 201n; Gio Antonio de Paoli, *The Lives of All the Roman Emperors* (N. and I. Okes, 1636), 289; William Prynne, *Historiarchos, or, The Exact Recorder Being the Most Faithfull Remembrancer of the Most Remarkable Transactions of Estate and of All the English Lawes* (Printed for Francis Coles, 1659), 64; Fuller, *The History of the Worthies of England*, 304; James Heath, *Englands Chronicle, or, The Lives & Reigns of the Kings and Queens from the Time of Julius Cæsar to the Present Reign of K. William and Q. Mary Containing the Remarkable Transactions and Revolutions in Peace and War, Both at Home and Abroad, as They Relate to This Kingdom, with the Wars, Policies, Religion and Customs, Success and Misfortunes as Well of the Ancient Britains, as Roman, Saxon, Danish, and Norman Conquerors* (Printed for Benj. Crayle, N. Bodington and G. Conyer, 1689), 45; James Tyrell, *The Generall History of England, as Well Ecclesiastical as Civil* (1696) 269; Spelman, *The Life of Ælfred the Great*, 34–35.

In steed of beautie a disfigurd face.
Our sights shall yeeld them: as you see me doe
(with that shee drawes a Knife) Virgins doe you.
Our bodies hetherto haue bene kept chast,
And unto death shall not this purenesse last?
Our bodies yet are free from foule lusts staine,
And shall we now be ravisht by the Dane.
Shall we polluted be with Pagans rape?
No no first perish this well-pleasing shape.
With streames of bloud w'eele quench unlawful fires,
With uglie lookes w'eele scarre untam'de desires.
Our spouse is JESUS, faith to him wee gave,
Hee shall our bodies chaste, though mangled have,
And though we be exteriorly foule,
He more will love the beautie of our soule.
In speech of men Euphrasia alone
Shall not hereafter live: of us each one
Shall acte that Virgin, and not feare deaths blow
That to our spouse unspotted wee may goe.
Empresse of Virgins, of our sexe the best,
To thee wee consecrate our snowy brest.
If any faint doe thou stout thoughts inspire,
JESUS pure Mother, give a noble fire.[65]

Notably, Ebba here justifies her 'murdring hand' by arguing that the Christian injunction against self-murder is cancelled out in this case by the moral necessity of giving oneself 'unspotted' to Jesus. That doubleness is repeated throughout the speech—the nuns are rebellious but obedient to God, 'exteriorly foule' but inwardly beautiful. The poem's narrator, however, sees things in a less nuanced way. They not only condone the mutilation as an act of 'happie Virgins'[66] but argue that the nuns' sacrifice was morally greater than that of the famed Lucretia, *because* they acted to prevent rape rather than to grieve or remedy it:

[65] The Euphrasia referenced here is Saint Euphrasia of Constantinople, who in the fourth century CE refused to marry a suitor favoured for her by the Emperor and instead joined a nunnery. Her presence here significantly links Ebba and the nuns to a tradition of monasticism as an assertion of self-determination which is also a turning-over of oneself to God's will.

[66] John Abbot, *Jesus Præfigured, or, A Poëme of the Holy Name of Iesus in Five Bookes* (1623), 80.

> Had she [Lucretia] before the acte with whetted knife
> Sundred in two the gold thread of her life,
> Wee would haue giu'n her a chaste Matrons praise.
> And wiser times her monuments should raise.[67]

This text frankly acknowledges that the nuns' actions are comparable to suicide, yet the sin of self-murder is overshadowed by the requirement for chastity. That message is underlined by the reference to 'Euphrasia', a virgin saint who escaped rape by her cleverness. A renowned beauty who was given to a barbarian after refusing to renounce her Christianity, Euphrasia promised him that if her virginity was preserved, she would give him a special herb which protected the user from enemy weapons, but only when received from a virgin. Having picked the herb, she offered to demonstrate its effectiveness by placing the plant on her neck and inviting the barbarian to give her a blow with a sword, which killed her. In arranging her own decapitation, Euphrasia safeguarded her chastity. As Euphrasia's story intimates, privileging of virginity may be at once misogynistic and a form of freedom, a way 'By bodies maime to set the body free'. Abbot himself recognised the parallels between the scenario he so vividly pictured and the gynocentric society of Amazons, similarly obedient to a matriarch, and similarly prepared to subject their bodies to 'noble cruelty' for a cause. In fact, he suggested, Ebba and her nuns ought to be considered '*Religious Amazons*':

> The Nuns who in obedience weare exact
> Follow their Abbesse in this worthy fact.
> Their Vizages (ô nobly cruel deed!)
> With plenteous streames issuing from wounds do bleed.
> Faces, where beautie dwelt, and eu'ry grace,
> (Religious Amazons) themselues deface.

Jesus Prefigured expands and embellishes the Ebba story. In so doing, it demonstrates how stories of Amazonian mastectomy provided some contemporary authors with an alternative interpretive framework for Ebba's self-mutilation, imagining it as a practical act of defence by a gynocentric community. This interpretation did not replace self-mortification, martyrdom, or asceticism as a context for the well-worn tale. Rather, it

[67] Abbot, *Jesus Præfigured*, 80.

supplemented and reinvigorated the issues of self-sufficiency and agency raised in those discussions, ensuring that Ebba's story remained alive for further generations of readers.

6.5 Death Follows Self-Injury

I have argued that in the seventeenth century, Ebba's self-injury could be read in new contexts and against or alongside new kinds of text. This intertextuality raised questions about what kind of a story this was, but also helped to renew its relevance and make it readable as a tale in which extreme bodily self-control allowed one to resist the violation of one's spiritual or psychological integrity. Nonetheless, neither modern nor early modern readers could see Ebba's self-mutilation as a straightforward triumph for female agency. While I have argued that Ebba shared much with the gynocentric Amazon society—including the use of self-injury as a means to bond an all-female community—the irony remains that Ebba's sisters were united in duty to their heavenly husband, Jesus Christ. Their protection of their virginity was thus as much an expression of that 'wifely' faithfulness as it was of their bodily autonomy, and this was recognised and praised in numerous accounts of the ordeal as an example of 'chastity and fortitude'.[68] Moreover, while these stories decried the barbarity of the invaders, they also located the 'wondrous beaut[y]' of the nuns as a factor in their likely rape.[69] Several texts even explicitly suggested that the nuns' beauty was a 'bait' or 'tentation' to their attackers.[70] Ebba and her sisters were therefore embedded in a culture which blamed women for their own rape, and they were presented as arguably having a moral obligation to mutilate themselves; one whose fulfilment was rewarded by being burned to death inside their home. Every version of the Ebba story concluded with this event, which reflects a seeming rule in early modern writing that sooner or later, 'defiled' women had to die. As noted at the beginning of this chapter, academic sources advised that rape victims should not end

[68] Richard Baker, *A Chronicle of the Kings of England, from the Time of the Romans Goverment unto the Raigne of Our Soveraigne Lord, King Charles Containing All Passages of State or Church, with All Other Observations Proper for a Chronicle* (Printed for Daniel Frere, 1643), 11.

[69] Defoe, *A Tour Thro' the Whole Island*, 8.

[70] Fuller, *The History of the Worthies of England*, 304; Louis Moréri, *The Great Historical, Geographical and Poetical Dictionary Being A Curious Miscellany of Sacred and Prophane History* (Printed for Henry Rhodes, Luke Meredick and John Harris, 1694), n.p.

their lives out of a misplaced idea of virtue, and Augustine even suggested that to do so displayed an attachment to honour that was prideful and therefore sinful.[71] However, as I have discussed in relation to Christian and Stoic discourses of suicide, the 'official', religious opposition to this practice and its representation in imaginative texts were seas apart. Just as the suicides of Brutus or Hieronimo might be represented as (albeit ambiguously) virtuous according to neo-Stoic values, so death—often by suicide—was continually represented as not only the most honourable, but the *only* honourable outcome for women who had survived rape. Thus, Lucrece/Lucretia served as a pattern for other women on stage or page who died either at their own hand, or in a 'mercy' killing like that of Shakespeare's Lavinia. The death of the rape victim was tragic, but it was also highly aestheticised, a variation on the grim fascination with violence against the female body which pervaded much of the period's literature.[72]

What has this to do with Ebba, whose actions explicitly *prevented* her rape? A clue may be found once again in representations of Amazon monomasty in early modern texts. Monomasty was, as I describe above, a matter of fascination for early modern ethnographers and their audiences. However, as I have argued elsewhere, the one-breasted woman is also a figure who must be kept at a distance, to the extent that she cannot even be directly represented on stage.[73] Plays about Amazons revelled in the gynocentric status of this tribe, but consistently obfuscated their bodily difference. In Mayne's *The Amorous Warre*, for example, a group of men are tricked into bed with women they believe to be Amazons. The next day, they each confirm that their partner *'had breasts'*—unaware that they have been sleeping with their own wives.[74] Moreover, medical writers, who wrote expansively on operations for breast cancer, never drew parallels between one-breasted survivors of mastectomy operations and Amazons. Instead, women who had lost a breast simply disappeared from the textual record. As Kathryn Schwarz has argued, the Amazon might be good to think with for a while, but ultimately constituted a threat to society that had to be quelled: 'Bringing them close to home invites chaos: as

[71] John Bugbee, "Chaucer's *Lucretia* and What Augustine Really Said About Rape: Two Reconsiderations". *Traditio* 74 (2019): 335–73, (337).

[72] On this trope, see Drew Daniel, *Joy of the Worm: Suicide and Pleasure in Early Modern English Literature*, Thinking Literature (University of Chicago Press, 2022), 43.

[73] Skuse, *Surgery and Selfhood in Early Modern England*, 35–54.

[74] Jasper Mayne, *The Amorous Warre* (S.N., 1648), pt. 5.1.

separatists they are a threat, but as mothers and lovers and wives and queens they are a disaster, participating in and altering the structures that should work to keep them out'.[75] The only 'tame' Amazons in early modern literature are those whose bodies are unaltered, and who can be accommodated within norms of heterosexual desire.

The textual erasure of one-breasted women and the idealised suicides of rape victims display the hallmarks of abjection. These kinds of bodies are equivalent to the 'refuse and corpses' which Julia Kristeva argues '*show me* what I permanently thrust aside in order to live'; they display the vulnerability, contingency and 'defilement' inherent in living.[76] In addition, however, these erasures also highlight either an unwillingness or an inability among early modern authors—and perhaps, among readers—to imagine a social place for women who were no longer valuable as objects of desire and exchange. This difficulty can also be seen in accounts of Ebba, which emphasise the extent to which her self-mutilation leaves her (and the other nuns) utterly 'disfigured', 'deformed' and even 'monstrous'.[77] Moreover, like those other 'unimaginable' women, the story of Ebba's self-mutilation shows how different kinds of bodily agency and anomaly might overlap to create a figure which is unthinkable in more ways than simply being undesirable. That is, Ebba's convent sisters become unattractive through their self-mutilation, but that mutilation *does more* by being emblematic of the way in which forcibly asserting one's rights over the reproductive body is itself a form of bodily heresy which cannot be allowed to stand. Wishing to avoid rape is a perfectly virtuous and legitimate desire. Actually managing to do so through the exertion of supreme bodily self-control is quite a different matter, and denotes a degree of personal autonomy which is hard to reconcile with a society in which women are ideally submissive and passive.

It is partly for this reason that Ebba and her sisters target their noses and not (for example) eyes, or breasts. Noses, as many contemporary

[75] Kathryn Schwarz, *Tough Love: Amazon Encounters in the English Renaissance* (Duke University Press, 2000), 23.

[76] Julia Kristeva, *Powers of Horror: An Essay on Abjection,* transl. by Leon S. Roudiez (Columbia University Press, 1982), 8.

[77] John Bale, *The First Two Partes of the Actes or Unchast Examples of the Englysh Votaryes Gathered out of Their Owne Legenades and Chronycles by Johan Bale* (1551), 52; Wilson, *The English Martyrologe*, 231; Crooke, *Mikrokosmographia*, 649; Prynne, *Historiarchos*, 64; Fuller, *The History of the Worthies of England*, 304; Heath, *Englands Chronicle*, 45; Baker, *A Chronicle*, 11.

authors observed, were regarded as the most essential part of the beauty of the face as well as a physiognomic indicator of character. In *Anthropometamorphosis*, for instance, Bulwer waxed lyrical on the 'Great … Ornament, that the Face receiveth by the nose … the imperiall seat of Majestique beauty'.[78] He went on to remark upon how Ebba's mutilation actually served as further evidence for the beauty and importance of the nose:

> that admirable stratagem of … the Abbas Ebba, and those Chast Virgins, which were in our Kingdome at the time of the Incursion of the Danes, who cut off their own Noses to preserve their Chastity, were both built upon this concession, that the Nose any way mangled or cut off, gives the greatest blemish to the Face, and proves most destructive to the enchanting beauty thereof.[79]

Bulwer's discussion acknowledges that the loss of the nose was a watchword for disfigurement. It was also a matter of reputation. Ebba's facial difference is described in Persons's influential account as 'shameful', and as numerous scholars have pointed out, lack of the nose stood powerfully for social, and particularly sexual, shame over hundreds of years.[80] In her work on medieval nose-cutting, for instance, Skinner notes that mutilation of the nose was a punishment which was much threatened but seldom employed, being perceived as excessive. Indeed, she finds that the only instance of nose-cutting in English texts before the eleventh century is in the *Life of St Swithun*, where the saint heals an innocent victim of this punishment.[81] In both European and Arabic stories in which women lost their noses, there was, she argues, a clear justification: 'women who transgressed deserved a permanent and disfiguring sign on their faces, and it seems that this sign transcended geographical and temporal boundaries'.[82] In his analysis of the subject, Valentin Groebner argues that late-medieval facial mutilation was used by individuals in retaliation for sexual infidelity

[78] Bulwer, *Anthropometamorphosis*, 116.
[79] Bulwer, *Anthropometamorphosis*, 118.
[80] Persons/Parsons, *An Examen*, 47.
[81] Patricia Skinner, "The Gendered Nose and its Lack: 'Medieval' Nose-Cutting and its Modern Manifestations". *Journal of Women's History* 26, no. 1 (2014): 45–67, (53).
[82] Skinner, "The Gendered Nose", 56.

or defamation, and was designed to shame the victim.[83] Most recently, in her comprehensive work on rhinoplasty in early modern England, Emily Cock shows how the procedure was vexed by moral considerations about restoring noses which had been lost to syphilis, and which ought to be readable as such. In response, authors emphasised cases in which noses had been lost in battle, or even in the process of resisting rape:

> The paragon story of the early seventeenth century … centres on the Duke of Savoy's attack on Geneva in 1590, wherein, 'a Virgin fell into the hands of the Soldiers, whose Chastity when they had attempted in vain, they being enraged cut her Nose off'. Two years later, her nose is restored 'so artificially, that, to the Admiration of all, it appeared rather Natural than Artificial' and we are told that she 'continues unmarried' in Lausanne at the time of writing (1613).[84]

This story makes for interesting reading in the light of the Ebba myth. Like Ebba, this virgin pays for avoiding rape with the loss of her nose. However, in this case, it is male rather than female agency which shapes and reshapes the imperilled body. It is unclear why the soldiers' attempts are 'in vain', but it is the decision of male aggressors to disfigure their victim and of a male surgeon to restore her to 'normalcy'. The detail that this virgin remains unmarried also seems significant. Does the attempted rape, or artificiality of the restored face, render this virgin unmarriageable? Or, is celibacy a continued choice which confirms the virgin's deserving status? Either way, avoiding rape, being disfigured and being restored are all events which act on an individual body, but affect a larger community.[85]

[83] Valentin Groebner, "Losing Face, Saving Face: Noses and Honour in the Late Medieval Town". *History Workshop Journal* 40, no. 1 (1995): 1–15. https://doi.org/10.1093/hwj/40.1.1.

[84] Emily Cock, "Exchanging Flesh: Prostitution and Plastic Surgery in Seventeenth-Century England" (PhD diss., University of Adelaide, 2013), 215.

[85] Another interesting story to consider here is that of Saint Brigit. After praying to be spared marriage so that she might remain celibate, one of Brigit's eyes spontaneously falls out of her head 'and by this meanes she became soe defformed, that noe man sought after her anie more'. She then enters a monastery, after which her eye is miraculously restored because 'her deare spouse [Jesus] would not endure, that she who had desired to loose her beautie to preserve her virginitie, should remayne in such deformitie'. Jerome Porter, *The Flowers of the Liues of the Most Renowned Saincts of the Three Kingdoms England Scotland, and Ireland Written and Collected out of the Best Authours and Manuscripts of Our Nation, and Distributed According to Their Feasts in the Calendar* (1632), 119.

Losing the nose is thus not only a bid to preserve virtue that looks like a punishment for sin (in much the same way that rape victims' suicide is a sin that looks like virtue). It is an assertion of moral integrity that, when exercised by women, must be accompanied by annihilation. In this way, nose mutilation might be considered as equivalent to the acts of self-castration examined above. After all, the nose and the phallus were conceptually linked.[86] Both Ebba's self-mutilation and that of self-castrating men were characterised by fear of a loss of sexual control, a motivation which makes self-injury comical in the case of castration ballads and tragic in stories of Ebba's martyrdom. However, where castrated men commit social suicide, self-mutilated women almost invariably end up dying for real. Like the rape victim, the mutilated woman is a powerful emblem of both abjection and endurance, but she cannot be allowed to persist, lest she flesh out an identity that is neither whorish, virginal, married or widowed.

6.6 Conclusion

Ebba's self-mutilation and that of her nuns was a horrifying story, but it remained popular throughout the early modern period because at every point, it spoke to contemporary questions about identity and integrity in terms that were relevant to its audience. Viewed in one way, Ebba had much in common with self-mortifying flagellants or self-starved medieval maidens. From another angle, Ebba and her sisters seemed more akin to Amazons in their willingness to disavow their beauty. While the story of Ebba may have changed little over 800 years, what it meant for readers to learn the story of Ebba might alter with the times. This flexibility depended in large part on the enduring ambiguity about what *kind* of story this was—martyrdom, tragedy or something else. By remaining unfixed, the account remained able to fulfil multiple functions. Ebba's ambivalent hagiography also highlights how enmeshed ideas of martyrdom were with early modern ideas of self-injury, morality and personal integrity. Even when Protestant readers rejected the idea of Ebba as worthy of sainthood, the idea of protecting one's ideological and spiritual self through the voluntary loss of physical integrity retained its power.

Ebba's tale was not included in the public work I shall describe in Chap. 8, for what are hopefully obvious reasons. Yet, her story as it was told and

[86] Groebner, "Losing Face", 5.

retold over the sixteenth and seventeenth centuries unites many aspects of narratives about self-injury in this period. Like Shakespeare's Portia, Ebba is a gifted rhetorician who understands that actions may speak louder than words. In common with Hieronimo, she carefully frames her self-mutilation as an act of self-sacrifice, but cannot easily escape the moral proximity between this act and the sin of suicide. Narratives about Ebba highlighted her beauty prior to mutilation, just as stories about violent lament often used the injuries inflicted therein as an opportunity to linger on the sexuality of the lamenter. Finally, Ebba was, like the self-castrating man, effectively erased by her self-injury—though unlike those men, Ebba's erasure was not merely symbolic and social but existential.

CHAPTER 7

Conclusion

On 18 March 1741, the Ordinary of Newgate made an account of how the people who were to be executed that day had conducted themselves whilst in prison. Those of whom the Ordinary approved had lived quietly, expressed repentance for their sins and attended religious services. The bad apples of the jail had either been sick when those services were taking place or had refused to engage with them, and denied or defended their crimes.

Among those the Ordinary described was a man named Daniel Jackson, who had been jailed for the murder of his wife the previous October. Jackson, the court heard, was a layabout and a drunkard who had unsuccessfully endeavoured to lead his wife down the same path. He psychologically and physically abused his wife, and finally, on 28 October 1740, shot her with a pistol as they were both exiting a coach. He was swiftly apprehended, tried and condemned to death. What happened next was recorded by the Ordinary in a lengthy text:

> On Sunday the 14th of December, while he was at Chapel, Charles, one of the Men who looks after the condemn'd Criminals, going into [Jackson's] Cell, found some Bloody-Rags, which gave him a strong Suspicion that Jackson had done himself some Mischief, whereupon he acquainted a Gentleman who was a Prisoner in the Press Yard therewith, and desired him to examine him about it, as he came down, [from] Chapel. Accordingly he did so, and put it very Home to him, remonstrating to him, how dreadful it

A. Skuse, *Hurt Feelings*, Palgrave Studies in Literature, Science and Medicine, https://doi.org/10.1007/978-3-032-04565-2_7

would be to add the Sin of Suicide, to that of the heinous Murder which he had already perpetrated.

Jackson at first denied it very obstinately, and on the Gentleman's telling him of the bloody Rags found in his Cell by Charles, said the Reason thereof was, he was troubled with the bloody Flux. However the Gentleman still pressing him yet more earnestly, to acknowledge the Truth, he condescended at last, tho' with great Difficulty, to confess that he had stabb'd himself in the Belly, and pulling up his Shirt, shew'd them the Place. Hereupon a Surgeon was immediately sent for, who having prob'd and dressed his Wound, enquired when, in what Manner, and with what Instrument he had made such a large Gash? His Answer was, that he did it the very Night he was committed to Newgate, being on Friday the 28th of November, between Eight and Nine in the Evening, with a Shoomaker's Knife, which he afterwards threw over the Hatch belonging to the Old Condemn'd-Hold. As to the Manner of doing it, he said it was as follows; whilst he sat on the Stairs of the Condemn'd-Hold, he took the Knife with both his Hands extended, and thrust it aslant, with all his Strength, which divided the Umbilical Vessels.

Accordingly on searching his Cell, the second Time, a large Lump of proud Flesh was found, which he had extirpated from the Navel, being a Sort of Fungous or Excressency, proceeding from the aforesaid Cut, and occasioned by the Wound's not having proper Care taken thereof, immediately after its being given. Hereupon the Surgeon ask'd, how he came to cut off that Part? To which he answered, it was so weighty and troublesome, he could not bear it any longer; adding, that it stunk to that Degree, and was so very offensive, it was intollerable.

He was then ask'd, if he expected to have saved his Life, by cutting it off; or, whether he thought it would have put him out of Pain: He reply'd, No, but that he did it to ease himself, and not to destroy himself; as also, that, if it pleased God to spare his Life, he never would attempt the like again; being willing to be made a public Example, that it might deter others for the future, from committing Murder, a Crime too frequently perpetrated in this City, and for which he was very justly condemn'd.[1]

When his warders first approached him, Jackson was insistent that his acts of self-injury (first stabbing himself, later cutting away the 'fungous' tissue that resulted) were not suicide attempts. In particular, his second act was undertaken '*to ease himself, and not to destroy himself*'. The intended

[1] *Old Bailey Proceedings Online* (www.oldbaileyonline.org, version 9.0) *Ordinary of Newgate's Account* September 1731 (OA17310924). https://www.oldbaileyonline.org/record/OA17310924 (Accessed: 13th June 2025).

purpose of his self-stabbing was obscure, but his narrative shared features with some of the accounts of self-injury which this book has examined. Jackson's first act was theatrical in character; in a scene strikingly reminiscent of Portia's self-stabbing, he too did the deed in private, but later recalled it in vivid detail and attempted to use it as the basis for a moral argument. The second attempt was meant to 'ease' Jackson, but not by alleviating pain. Instead, his description of excising a 'troublesome', 'offensive' piece of flesh speaks to the protagonist's continued ability to exercise control over his own body despite incarceration, a mixture of do-it-yourself surgery and curious symbolism. The bloody rags and the hidden discarded flesh cut from the 'umbilical vessels' bring to mind an aborted foetus, while Jackson's bizarre decision that cutting off a lump of his own tissue would be preferable to living with an 'Excressency' suggests that the flesh had acquired metaphorical meaning. Did this 'proud flesh' stand for an aspect of himself which he rejected, as in seventeenth-century tales of self-gelding? Or, was its excision meant to communicate a message to other people, as in Hieronimo's autoglossotomy? It is clear that whatever his original intention, when he was confronted Jackson optimistically surmised that his self-inflicted wound might do communicative work by making him a 'public example'. Thus, he hoped, this act of apparent self-destruction might actually prolong his life.

In the event, Jackson was unable to turn his self-injury to the uses he envisaged. The Ordinary recounted:

> He did not live long after making this Discovery, for on Monday the 15th of December, Abraham Mendez (a very faithful and trusty old Servant belonging to the Gaol) going about Eight in the Morning, being the usual Hour to unlock the Cells of the Condemn'd Criminals, and see whether they wanted any Thing, when he came to that of Daniel Jackson, he desired him to fetch him a Glass of warm Wine, for he was very faint, and prodigiously indisposed; adding, that if he did not make a great deal of Haste with it, he should certainly die; wherein he proved a true Prophet, for before Abraham return'd with it, he gave up the Ghost.[2]

On this scanty evidence, Jackson appears to have died of an acute sickness, likely an infection resulting from his injury and the filthy conditions of the jail. Notably, the Ordinary wrote that he believed from the above

[2] *Ordinary of Newgate's Account* September 1731.

circumstances that the prisoner 'committed a premeditated Self Murder' despite Jackson's protestations to the contrary. He was soon vindicated by the discovery of a letter addressed to the prisoner's mother:

> DEAR MOTHER,
>
> THERE is one Thing I beg and crave of you, that is, you make all the Interest you possibly can, leave no Stone unturn'd to grant me the Happiness of being buried with my dear Wife, though in ever so mean a Manner, the meaner the better to my Satisfaction, if it is without a Coffin I am content. If it had pleased God I had got off at my Trial, it was my full Intent to have buried myself with her, as soon as I could find out her Grave, otherwise I had pleaded Guilty, and not have given the Trouble of Trying me. I shall die this Night with as much Pleasure as I went to be married, being satisfied within myself God Almighty will forgive me my great and manifold Sins, through the Merits of Jesus Christ. I hope none will reflect on my Behaviour on seeming unconcerned before Company, my Heart was not so, I could cry at others Misfortunes, but hated myself for such an inhuman Crime.
>
> Dear Mother and Sisters, make yourselves easy, I doubt not, through God's Mercy, I am happier than this World could make me. The World will blame me for Self Murder; but after reasoning with myself, that I had forfeited my Life, that Blood cries for Blood, I could not think it any Offence to Almighty God, to pay the Debt myself I ow'd my dear Wife; I having more Time than when in a Crowd, to prepare myself for Eternity.
>
> Pray let the Ordinary of Newgate have a Copy of this Letter; I have left him the Particulars, relating to this unhappy Scene, in another Paper.
>
> Adieu! dear MOTHER, and SISTERS;
>
> DANIEL JACKSON[3]

When Jackson wrote the letter, he clearly knew he was dying (in addition to the fact that he was condemned to be executed). It does not necessarily follow that he had injured himself several weeks earlier with suicidal intent. While his letter talked of wishing to follow his wife into the grave, this was not something which he had mentioned when the wound was first discovered, and indeed he initially denied any such plan. Indeed, the fact that Jackson, in prolonged possession of a knife, chose to follow up his

[3] *Ordinary of Newgate's Account* September 1731.

self-stabbing by cutting off and hiding a piece of his flesh rather than more expediently slitting his throat, suggests that suicide was *not* his primary intention. By contrast, William Bird, who was jailed in 1740 for having killed his wife, stabbed himself '6 or 7 Inches into his Body' with a sword as soon as he realised his wife was dead.[4] When Bird was asked how he had done such a thing, he asked for his sword back so that he could demonstrate—unlike Hieronimo's hapless captors, the attending surgeon wisely rejected this request. Robert Irwin, who awaited execution for murder in 1721, stated that he wished to die and was upbraided for that desire by the Ordinary. Nevertheless, 'he fasted much, and said he deserved not to eat'.[5] The fact that the Ordinary readily and insistently interpreted Jackson's actions as suicidal despite obvious differences from other attempted suicides by prisoners thus suggests a possible change by this period in how self-injurious acts *worked*, symbolically speaking. Incidents with the hallmarks of attempted suicide were lumped in with phenomena in which death was clearly not the intention. The authority to determine how an injury was judged no longer rested with the person injuring themselves but with the onlooker.

* * *

Jackson's is just one case, but some examples from eighteenth-century literature similarly, and intriguingly, suggest that even as the cult of sensibility came into being, attitudes towards self-wounding were shifting away from sympathy and towards hostility or even ridicule. For instance, Letitia Pilkington's *Memoirs*, published in 1748, told how the future Mr. Pilkington reacted when they were set to be separated by her parents:

> [I] thought proper to give Mr. *P____n* Notice of my Departure, and easily prevailed on my Brother to give him a Letter; but Heav'ns! How was I frighted when he, returning in few Moments, told me, Mr. *P____n* had stabbed himself with his Penknife: I ran all in Tears to my Mother, entreating her Permission to go and see him. She appeared much concerned, and sent for him to come to us, which pleasing Summons he readily obeyed. But I could scarce forbear laughing at my own Credulity, when my wounded

[4] *Old Bailey Proceedings Online* (www.oldbaileyonline.org, version 9.0) *Ordinary of Newgate's Account* March 1741 (OA17410318). https://www.oldbaileyonline.org/record/OA17410318 (Accessed: 13th June 2025).

[5] *Old Bailey Proceedings Online* (www.oldbaileyonline.org, version 9.0) *Ordinary of Newgate's Account* April 1721 (OA17210403). https://www.oldbaileyonline.org/record/OA17210403 (Accessed: 13th June 2025).

> swain came to us in perfect Health. He had indeed given himself a Scratch, on Purpose to terrify us, and had just such a desperate Wound as I have frequently received from the Point of a Pin, without complaining.[6]

Mr. Pilkington makes a variety of the same error committed by Tamburlaine. He is invested in the symbolic power of self-injury, but fails to back it up with the physical pain that gives such an act meaning. By this point, however, it is a piece of 'credulity' to believe that such bodily peril *would* really exist. A similar scenario was played out in the 1749 novella *The ladies advocate: or, wit and beauty a match for treachery and inconstancy*. In this tale, the over-eager paramour of 'Philly' attempts to stab himself when his suit is thwarted, and later repeatedly threatens suicide if Philly attempts to leave him.[7] The story was partly based on Pilkington's life, as well as that of the notorious Teresia Constantia Philips, and firmly placed self-injury as emotional blackmail. Moreover, this judgement could still hold when the injury was more consequential, as in a story sent by Lady Manley Delarivier to Alexander Pope in 1735 (she had apparently heard it from the gentleman concerned at a dinner party; a man who married a wife of considerable fortune which he later attempted to attain):

> [After the wedding] Three happy Months I had all the Satisfaction that innocent Marriage and excessive Love in a Bride, could give me. Then I began to consider a little my Affairs, and proposed to my Wife my being naturalized, that I might look after hers. She swooned at the Name; and when she recovered, she snatched a Bayonet of mine, and wounded herself under the Left Breast, but not much. I cannot express my Surprize: We hushed the Matter for fear of her Mother; and I employed some of my Soldiery Skill to cure it, which had the [desired] effect. I enquired into the Reason of this Extravagancy. She told me the Discovery *of interest* in me, when she had believed *Love* was the only motive of *our Marriage*.[8]

The wound in this case was less trifling, but the disapproval was the same. Self-injury was now evidence of 'Extravagancy' rather than sincerity. The

[6] Laetitia Pilkington, *Memoirs of Mrs. Lætitia Pilkington, Wife to the Rev. Mr. Matthew Pilkington* (1748), 1:35.

[7] *The Ladies Advocate: Or, Wit and Beauty a Match for Treachery and Inconstancy* (Printed for C. Long, 1749), 195–96.

[8] Alexander Pope, *Mr. Pope's Literary Correspondence, Volume the Third. With Letters To, and From, the Duke of Shrewberry, Dean Swift, Lord Lansdowne, Lady Chudleigh, Bishop of St. Asaph, Mrs. Manley, Sir Berkeley Lucy, Mrs Thomas, &c.* (E. Curll, 1735), 74.

capacity of self-wounding to garner attention has been over-exploited, and is diminished as a result.

We might usefully contrast these examples with two cases in Hugh Ryder's *New practical observations in surgery*, published in 1685. Ryder recounted a number of instances in which self-injury followed interpersonal conflict. The first was a story of a young soldier, recently returned from Europe:

> upon some disgust he had received from his Mother, with a large Flemish knife stabb'd himself about two inches above, and on one side of the Navel, through the right Muscle and *Peritonaeum* into the *Abdomen;* from out the wound hung forth a large piece of the *Omentum,* which had so done for four hours, for so long after it was that I saw him: it being Winter time, the Air having seiz'd it, and being discoloured, I made ligature upon it, and cut off the discolour'd part near the the tying, and having fomented it with warm Brandy, I put in the *Omentum* again, leaving the thread hanging forth to the wound; I used a Digestive for three or four days, afterwards a Balsam, whose basis was Turpentine; so without any supervening symptom he became well in less than a fortnight.[9]

The second tale was of a shoemaker's wife, who

> her Husband not presently granting her unreasonable demand, took one of his sharpest Knives wherewith he used to work, and stabb'd her self obliquely into the *Epigastrium:* the Wound was small, (and so was the Knife, which I saw was Bloody about two Inches or more) and it bled but little; wherefore I took Blood from her to prevent inward bleeding which I feared; I drest the Wound with a short Tent with a Digestive. There hapned after 2. days a large tumour round about the wounded place, of which she complained very much, and proved to be only from wind; for upon the use of warm Fomentations, it was wholly discuss'd, and she in seven or eight days became well.

Ryder's first patient gave himself an injury markedly similar to that of the condemned prisoner described above. His second responded violently and impulsively to a perceived slight in the same manner as the woman described by Lady Delarivier. Yet, the way in which these instances are described differs radically from those later examples. Ryder does not

[9] Hugh Ryder, *New practical observations in surgery containing divers remarkable cases and cures* (James Partridge, 1685), 24–25.

approve of self-injury, but he does not go out of his way to moralise against it, mock it, or assign probable causes for it. When his patients declare the cause of their actions, he believes them. Moreover, he does not surmise that the patients in question intended suicide. This is in part a function of genre; Ryder was a meticulous surgeon, with a keen interest in the healing of wounds. Nonetheless, when set against comparable examples from only decades later, the accounts are striking for their seriousness, and for centring the voice of the self-injuring person.

Taken together, these tales tentatively suggest that the window during which self-injury could be read as a valid means of self-expression began to close in the early eighteenth century.[10] Instead, interpretations of these acts now routinely occupied what had been the two extremes of their possible meaning—on one end, foolish and unserious attempts at attention-seeking, and on the other, sinful attempts at suicide. What was more, people who witnessed self-injurious acts insisted that *they* were the ones who should decide on the act's meaning, and the ability for a person who self-injured to communicate how their actions were to be interpreted by others was curtailed. It was a self-fulfilling prophecy; when self-injury was read as evidence of 'extravagancy', foolishness or degeneracy, then attempts by the protagonists to cast their actions in a different light were disregarded. They were, after all, extravagant, foolish or degenerate, and not to be credited. In work on pin-swallowing in early modern England, Imogen Knox notes a similar trajectory. Ingestion of objects such as pins, which was taken seriously as a supernatural or natural sign in the seventeenth century, later became read as evidence of either trickery or hysteria (Imogen Knox, 2023).

This eighteenth-century shift away from viewing self-injury as a legitimate means of communication remains a topic for further study, beyond the scope of this book. It preceded, and possibly paved the way for, the 'invention' of self-harm as an object of study in the nineteenth and twentieth centuries. Sarah Chaney, for instance, finds that in the nineteenth century self-mutilation, 'a new category of psychiatric symptomatology', was firmly associated with both madness and immorality. Through its use as a rhetorical device in novels, it was represented as 'the ultimate act of selfish preoccupation', showing a worrying disregard for the 'social

[10] This chapter in the representation of self-injury deserves further study, especially its links with the growing business of madhouses.

instinct' understood to underpin normal society.[11] This interpretation of self-injury as a pathology which was also morally freighted continued into the twentieth century. Chris Millard suggests that during this period self-harm was interpreted as 'a largely uncomplicated attempt to die' or later 'a pathological communication', with interpretations focussed on sociality and affect regulation only coming into play late in the century.[12] Such readings closed down possibilities imagined in the early modern period. The expansive, sociologically oriented readings of self-injury which we see in the next chapter are a recent phenomenon, reopening a field of interpretation previously narrowed to include only medical and moral perspectives.

* * *

Why, then, did the sixteenth and seventeenth centuries allow, and encourage, a richness and variety of interpretations of self-wounding which later centuries did not? As the preceding chapters have shown, the answer is linked to a raft of complex social and cultural factors, mostly having to do with the significance of pain. Pain in this period was *useable* to a greater extent than in subsequent centuries. Judicial corporal punishments told citizens that pain was a means to destroy the individual, while interrogatory torture sought to crack open a person to discover truths hidden within. This idea interacted in complex ways with religious representations of martyrdom and bodily sacrifice. Catholic hagiographies frequently depicted the ability to withstand pain as a form of necroresistance, which was also spiritually purifying. Mortification could be imagined as both an abnegation of the self, and an assertion of one's personal relationship with God. Though Protestantism ostensibly rejected these stories, it also rehearsed them through tales of religious martyrdom.

Neo-Stoicism also had an important part to play here; its relatively minor revival was magnified through the lens of popular drama. The precedence of classical literature in boys' education (and the consequent flush of neo-Senecan plays) meant that mass audiences were repeatedly exposed

[11] Sarah Chaney, "'A Hideous Torture on Himself': Madness and Self-Mutilation in Victorian Literature", *Journal of Medical Humanities* 32, no. 4 (2011): 279, https://doi.org/10.1007/s10912-011-9152-6; see also Sarah Chaney, *Psyche on the Skin: A History of Self-Harm* (Reaktion Books, 2017).

[12] Chris Millard, *A History of Self-Harm in Britain: A Genealogy of Cutting and Overdosing* (Palgrave Macmillan, 2015), 2.

both to notions of self-sacrifice as potentially noble, and violence itself as having aesthetic value (an idea which looped back to the spectacle of executions). On stage, violence was both glamourised and invested with political and moral purpose.

That the body assumed such symbolic and social importance was also due in part to the way in which it was understood. The doctrine of the humours inextricably linked physiology and psychology, so that inner life was reflected in the flesh and vice versa. This too was what made accounts of bodily difference, such as those from the New World, so intoxicating for early modern readers. Arguably, however, all these factors united in the centrality of rhetoric to sixteenth- and seventeenth-century culture. By including gesture as a central part of rhetoric, and insisting on the 'personation' of emotions by the ideal rhetor, the theory of rhetoric wound together expression and experience, emotional suffering and physical pain, symbolic form and somatic content. Discussions of rhetoric were always reaching for a way in which a rhetor's authenticity could be proven or disproven, and though this would ultimately prove elusive, self-wounding looked, at times, like a good way to demonstrate one's sincerity.

Altogether, these factors allowed self-wounding to be considered an act which *did things* for the protagonist. What those things did might be depended on the context. In plays which drew on neo-Stoic antecedents, self-injury was readily imagined as a supreme example of self-mastery. To hurt oneself was, in some cases, to declare the supremacy of the mind over the body. In others, it was to reclaim biopower; having sovereignty over oneself requires that all the body's capacity, including its capacity to be destroyed, remains within one's own control. Off-stage, poetry and polemic engaged with life's messier realities. Self-injury could be used as a way to manage the emotions, giving physical form to mental pain and thus making that pain visible to others. When it wrought permanent alteration, it might declare one's intention to exit systems which insisted that certain sorts of bodies automatically entered into relationships of subjection or dominance based on gender or status.

Most of these uses for self-injury were attention-seeking in some sense. Yet, that did not have the same pejorative connotations as today. Individuals sought to communicate what was happening 'inside'—distress, anger, determination, devotion—by marking their 'outside'. This was appropriate to a culture in which verbal rhetoric was vitally important but also distrusted, and where visible proof was of ever-increasing importance. That is not to say that for early modern people, self-injury was normal or welcome. Acts of self-wounding still shocked those who saw them, and this was part of their power; they demanded witnessing. Nonetheless, early modern responses to self-wounding tended to ask 'why?' rather than demanding 'stop'. This was not a problem in need of a solution.

The following chapter, the last in this book, looks at some recent sociologically focussed work on self-injury as a communicative phenomenon in the twentieth and twenty-first centuries. These readings, I argue, reopen a range of interpretive possibilities for self-wounding with striking similarities to the shifting, lively narratives I have described at work throughout the turbulent sixteenth and seventeenth centuries. At the same time, comparison between the modern and early modern risks flattening the complexity of both, and creative interventions into the topic highlight the risks and rewards of presentism, as well as fault lines in the notion of 'impact'. As Daniel Jackson and the Newgate Ordinary knew well, the meanings of self-wounding were often hard to puzzle out. However closely one might examine the deed, its significance was never fixed.

CHAPTER 8

Coda: Self-Injury, Anachronism and Engagement

PORTIA: *I think we need to stop seeing it as "attention-seeking". We are all seeking attention and we communicate this in different ways.*

Thus far, this book has suggested links between early modern and modern experiences of self-wounding without fully exploring that connection. Each chapter (barring the Introduction and Conclusion) is preceded by quotes from the play written as part of creative workshops related to the 'public' part of this project, and there are some more substantial points of connection; in Chap. 3, for instance, I briefly considered how the modern categorisation of some self-injury events as unconnected with suicidal ideation, and therefore less worthy of notice, compares with Renaissance representations of self-wounding as powerfully communicative. Throughout this project, however, the complexities of comparing actions which look the same but are understood differently have been keenly felt. This final chapter is therefore an unusual intervention in a primarily historicist text—an attempt to reflect on those challenges through theories about (un)historicism, self-injury and publicly engaged research, and via a description of my own imperfect practice.

The first part of this chapter examines the topic of anachronism broadly. It does so in part by considering the debate over 'Unhistoricism' which took place in the pages of the *PMLA* in 2013, prompted by thought-pieces from Madhavi Menon, Jonathan Goldberg and Valerie Traub. This debate has revolved around queer studies. However, I consider whether, and how, the principles and critiques of 'unhistoricism' might be

A. Skuse, *Hurt Feelings*, Palgrave Studies in Literature, Science and Medicine, https://doi.org/10.1007/978-3-032-04565-2_8

of use in medical humanities, particularly in medical history-inflected studies such as this one.

The second part of the chapter looks at how self-injury is theorised in the twenty-first century, and what those theories might share (or not share) with the stories in this book. Self-injury is now highly medicalised, but it has also been the subject of recent sociological studies which emphasise communication and self-regulation as motivating factors for people who self-injure. Are these areas of shared interest best understood as parallels between two separate phenomena (historical and present self-injury), or as evidence of what Menon and Goldberg would call a unifying 'homohistory'? I suggest that while histories of self-injury should not be homogenised, they may nonetheless help to 'queer' our understanding of modern self-harm by multiplying the interpretive possibilities for this practice.

The third part of the chapter suggests the importance and difficulty of bringing these theoretical debates to bear on research practice. I relate the public-facing events which formed part of this research project, in which people with lived experience of self-harm were invited to reflect creatively on the history of attitudes towards self-injury. This kind of research may help to close the perceived gap between past and present, but it is not without its issues. As I will discuss, over the course of these events, publicly engaged research turned to co-creation, and in so doing revealed much about how un-historicising research affects both academic and non-academic participants.

Finally, the fourth part of this chapter reflects on the lessons of these events. I show that participants identified striking comparisons between their own experiences and those of historical figures, which led to reflection on the biases surrounding self-injury. The notion of 'attention-seeking' provides a strong link across the centuries. However, the programme also highlighted the difficulties implicit in a University system which positions impact as the necessary end of engagement. Drawing on recent scholarship on the ethics of community engagement, I suggest that genuinely co-creative engagement requires a less pedagogical and more democratic approach which resists the temptation to treat co-creators and their experiences as subjects in need of improvement.

8.1 Anachronism Versus Unhistoricism

As I identified in the Introduction to this book, this research project consciously imposes a modern category, self-injury, onto a group of acts which did not share a terminology in the period of which I am writing. I have argued that disparate stories such as Portia's self-cutting and the jealous Quaker's self-gelding share narrative features and motivating logics. I hope to have shown that they exist in relationship with one another, though those relationships may be loose and sometimes unstable. What is less clear is what relationship this constellation of acts has (or does not have) with the modern phenomenon of non-suicidal self-injury (NSSI), or 'self-harm'. This question requires some specific attention to the phenomena under discussion, which I provide below. However, it is also involved with broader debates in literary and cultural theory, namely with how we approach the past. Is it true that, as the saying goes, 'the past is a different country; they do things differently there'? Or, does insistence on viewing past phenomena as bounded by their unique and contingent contexts reinforce an unhelpful binary between 'then' and 'now'?

These questions have informed the work of numerous scholars for almost a century. Oscar Moro-Abadia, for instance, names Herbert Butterfield and Hélène Metzger as having pioneered 'anti-presentism' in the 1930s, with this doctrine eventually having led to the emergence of New Historicism in the 1980s and its dominance as a mode of reading thereafter.[1] Since the 1990s, however, academics—and in particular, queer theorists—have increasingly engaged in what Jeremy Webster describes as 'contentious debate over whether people living in the past were fundamentally different from modern people or more or less the same'.[2] In particular, the debate has been focussed around whether, as Foucault suggested in his 1976 *History of Sexuality*, earlier periods of history knew 'sodomy' as a category of sexual activity but not orientation, with 'the homosexual' emerging as an identity category in the nineteenth century. In a series of thought pieces in the *PMLA* in 2013, 'unhistoricists' and historicists clashed. Jonathan Goldberg and Madhavi Menon argued

[1] Oscar Moro-Abadía, "Thinking about 'Presentism' from a Historian's Perspective: Herbert Butterfield and Hélène Metzger", *History of Science* 47, no. 1 (2009): 55–77, https://doi.org/10.1177/007327530904700104.

[2] Jeremy W. Webster, "Queering the Seventeenth Century: Historicism, Queer Theory, and Early Modern Literature," *Literature Compass* 5, no. 2 (2008): 378, https://doi.org/10.1111/j.1741-4113.2008.00527.x.

animatedly that historicist methods were usually, perhaps inescapably, teleological, and that teleology was fundamentally incompatible with queer ways of reading. Queer reading, they argued, rejected 'alterity': 'Why has it come to pass that we apprehend the past in the mode of difference? How has "history" come to equal "alterity"? And what effect does the privileging of the hetero have on studies of sexuality?'[3] Instead, they proposed a 'homohistory' that 'would be invested in suspending determinate sexual and chronological differences while expanding the possibilities of the non-hetero, with all its connotations of sameness, similarity, proximity, and anachronism'.[4] Importantly, this would not mean arguing that past sodomy was more similar to than different from present homosexuality, but rather jettisoning such definitions and 'resisting the strictures of knowability itself'.[5] Furthermore, such an approach would seek to deconstruct historical fields such as 'early modern', since they both divide the past into arbitrary segments and implicitly position it as having meaning chiefly in relation to the present.

These radical suggestions proved highly influential, and versions of unhistoricism (including some which precede Goldberg and Menon's piece) have been highlighted by Carla Freccero and Chris Roulston among others.[6] However, in a response to 'Queering Unhistoricism' published shortly afterwards, Valerie Traub contended that the unhistoricists' argument was lacking in substance. While they might repudiate bounded time periods, she argued, many of these scholars continued to write about 'the medieval' or 'the Renaissance', for the same reasons of utility that literary and historical scholars have always focussed on particular stretches of time. Moreover, the unhistoricists' insistence on deconstruction was, Traub argued, an overreach: 'Readings ... are not the same thing as history; and deconstructive and psychoanalytic interpretations of literary texts, while they contribute importantly to historical understanding, do not necessarily conduce to a historical explanation ... these critics overestimate

[3] Jonathan Goldberg and Madhavi Menon, "Queering History", *PMLA* 120, no. 5 (2005): 1608–17, (1609).

[4] Goldberg and Menon, "Queering History", 1609.

[5] Goldberg and Menon, "Queering History", 1609.

[6] Carlosa Freccero, Madhavi Menon, and Valerie Traub, "Historicism and Unhistoricism in Queer Studies", *PMLA* 128, no. 3 (2013): 781–86; Chris Roulston, 'New Approaches to the Queer 18th Century', *Literature Compass* 10, no. 10 (2013): 761–70, https://doi.org/10.1111/lic3.12103.

its analytic capacity and explanatory power'.[7] To equate the difference between sequential and non-sequential views of time with a series of other differences (homo/hetero; change/stasis; distance/proximity) was, she argued, a rhetorical act rather than a logical conclusion, and though time could be sliced and diced in various ways, including disruptive and subversive ways, time itself could not be dispensed with.[8]

These are battle lines; the most committed combatants are seemingly in search of an un-chronological way of doing history, and it is difficult to imagine how this might manifest. Subsequently, however, other scholars have suggested a less combative, and perhaps more workable approach, which tends to boil down to the conscious and careful use of anachronism or presentism as a tool for thought in certain circumstances. For example, Roland Betancourt, in thinking about medieval people and genderqueerness, has criticised 'hyperchronicity', an obsession with flat historical 'facts' that threatens to excise the 'human' from 'humanities'.[9] In Sanjay Seth's work on post-colonialism and sociology, he similarly speculates that *in certain circumstances*, anachronism can be 'redeemed':

> even where anachronism cannot be avoided, history writing can lead us to critically examine the continuities and breaks between the past under examination and our present. In some cases, such as in the preceding examples, we may … conclude that there is a bridge connecting 'us' to 'them,' assuring us that just as we are born of them, so our categories are also born of theirs; that the transformations and ruptures that make our present different from their past also arise out of that past, rather than being wholly external to it. The application of the code to the past may illuminate the emergence of the code; the 'former structurings and forgotten histories' that have brought us to our present historical situation can be disinterred and re-examined, so

[7] Valerie Traub, "The New Unhistoricism in Queer Studies", *PMLA* 128, no. 1 (2013): 21–39, (30).

[8] For their part, Menon replied sharply to Traub that unhistoricism was vital to 'talk back to the dominant modes of historicism', and that to assume unhistoricism would *not* yield new insights was 'baffling'. Madhavi Menon, "Historicism and Unhistoricism in Queer Studies", *PMLA* 128, no. 3 (2013): 782–84, https://doi.org/10.1632/S003081290012317X.

[9] Roland Betancourt, '"To Clarify an Everchanging Present"—Hyperchronicity, Anachronism and the relationship between history and social science' *LSE Impact Blog*, May 10th 2023, https://blogs.lse.ac.uk/impactofsocialsciences/2023/05/10/to-clarify-an-everchanging-present-hyperchronicity-anachronism-and-the-relationship-between-history-and-social-science/.

> that historical inquiry yields a better understanding of the tradition that produced us and through which we speak.[10]

In a chapter provocatively titled 'In Defence of Presentism', David Armitage presents a similar argument in greater depth. He outlines different varieties of presentism, ranging from 'ideological' presentism, which seeks to use history to bolster the superiority of 'modern' ways of thinking and being, to 'perspectival' presentism, only interested in the past insomuch as it directly informs the present.[11] For many historians, he notes, all these kinds of presentism are more or less problematic, but looking to other disciplines—including the history of science—can yield more benign forms. Drawing on two historians of science, Naomi Oreskes and Laurent Loison, Armitage cycles through still more types of presentism, this time less contentious, before finally landing on a version inspired by Augustine's conception of history. Since, as Augustine argues, we can only view any moment from our own perspective (a 'past present, present present, and future present'), there is a degree of presentism in every analysis. 'Human flourishing', concludes Armitage, 'is at once present-centred, future-oriented, and past-dependent'.[12]

This conclusion is one implicitly shared by many medical humanities scholars, whose business is to understand that the researcher—whether scientist, literary scholar, geographer, or anthropologist—is embedded within their own cultural context and that this unavoidably affects the kinds of questions they can ask. The health humanities are a particular case in point; most academics in this field can attest that even if they steadfastly refuse to draw connections between historical and modern medicine, any non-specialist reader will readily do the job for them.[13]

[10] Sanjay Seth, *Beyond Reason : Postcolonial Theory and the Social Sciences*, Oxford Scholarship Online. (Oxford University Press, 2021), 126.

[11] David Armitage, "In Defense of Presentism", in *History and Human Flourishing*, ed. Darrin M. McMahon (Oxford University Press, 2022), 50–53, https://doi.org/10.1093/oso/9780197625262.003.0003.

[12] Armitage, "In Defense of Presentism".

[13] When talking about early modern medicine to non-specialist audiences, the two questions I am most often asked are 'did any of it work?' and 'did they know what [diabetes/cancer/depression] was?'.

8.2 Self-Injury Sociologies

If we can accept, therefore, that a degree of anachronism is certainly likely, perhaps inevitable, or possibly even desirable, how might this function in the case of self-wounding? What, if anything, might contemporary and historical understandings lend to one another?

Modern discourses about non-suicidal self-injury tend to emphasise both the potential of this practice to serve as a 'cry for help' and the dangers it poses, including the link between self-injury and suicide.[14] Media coverage also focuses on the idea that self-harm is encouraged through social networks, particularly online. In August 2022, a private member's bill was introduced in the United Kingdom's House of Lords which would require the Office for Communications (OFCOM) to advise the government about the extent of content on social media platforms which might encourage people towards either self-harm or suicide (since 2023, the bill has been awaiting a third reading in the House of Lords, after which it would require approval by the House of Commons. Progress was halted by the general election in 2024, and at the time of publication, it seems unlikely that the bill will come to fruition).[15] News stories proliferate of self-harm being 'encouraged' online through various means, from people sharing details of their practice through coded social media hashtags, to malicious invocations to harm oneself as part of online bullying. 'Digital self-harm' is used to describe the practice of writing derogatory or hateful comments about oneself online. However, the level of concern about social media's role in self-harm is somewhat complicated by numerous studies which have been undertaken on the relationship between social

[14] Rebecca C. Groschwitz, Michael Kaess, Gloria Fischer, et al., "The Association of Non-Suicidal Self-Injury and Suicidal Behavior According to DSM-5 in Adolescent Psychiatric Inpatients", *Psychiatry Research* 228, no. 3 (2015): 454–61, https://doi.org/10.1016/j.psychres.2015.06.019; Janis Whitlock and Kerry L. Knox, "The Relationship Between Self-Injurious Behavior and Suicide in a Young Adult Population", *Archives of Pediatrics & Adolescent Medicine* 161, no. 7 (2007): 634–40, https://doi.org/10.1001/archpedi.161.7.634.

[15] Charley Coleman, "OFCOM (Duty Regarding Prevention of Serious Self-Harm and Suicide) Bill [HL]: HL Bill 18 of 2022-23". https://lordslibrary.parliament.uk/research-briefings/lln-2022-0041/. (Accessed 8 September, 2022).

media use and prevalence of self-harm, some of which show a positive and others a negative correlation.[16]

Despite the fact that concerns about online media implicitly recognise self-harm as a phenomenon subject to socio-cultural influence, the discourse about self-harm is primarily medical; it is deemed an indicator of psychological distress which may be pathological in nature. The latest edition of the *Diagnostic and Statistical Manual of Mental Disorders* (DSM-5), the standard classificatory text used by mental health professionals in the United States, introduces 'non-suicidal self-injury disorder' (NSSI) for the first time as a stand-alone condition. Its criteria are five-fold, including: the expectation that NSSI will solve an interpersonal problem, provide relief from unpleasant thoughts and/or emotions, or induce a positive emotional state; the experience of one or more of the following: (a) interpersonal problems or negative thoughts or emotions immediately prior to NSSI, (b) preoccupation with NSSI that is difficult to manage or (c) frequent thoughts about NSSI; that the NSSI is not socially sanctioned or restricted to minor self-injurious behaviours; the presence of NSSI-related clinically significant distress or interference across different domains of functioning (e.g. work, relationships).[17] More commonly used in the United Kingdom is the *International Classification of Diseases* (ICD-11). This text does not list a separate disorder of non-suicidal self-injury, categorising it instead as a 'symptom or sign' of various kinds of mental disorder including major depression, post-traumatic stress disorder and borderline personality disorder. The National Health Service website firmly positions self-injury as indexed to psychological factors, advising readers that reasons for self-injury include 'expressing or coping with emotional distress', 'trying to feel in control', 'a way of punishing themselves', 'relieving unbearable tension', 'a cry for help' and 'a response to

[16] Amber Barthorpe, Becky Mars, Paul Moran, and Lizzy Winstone. "Is Social Media Screen Time Really Associated with Poor Adolescent Mental Health? A Time Use Diary Study". *Journal of Affective Disorders* 274 (2020): 864–70. https://doi.org/10.1016/j.jad.2020.05.106; Nina H. Di Cara, Oliver S. P. Davis, Claire M. A. Haworth, Luke Sloan, and Lizzy Winstone. "The Mental Health and Well-Being Profile of Young Adults Using Social Media". *Npj Mental Health Research* 1, no. 1 (2002): 1–11. https://doi.org/10.1038/s44184-022-00011-w.

[17] Kim L. Gratz, Alexander L. Chapman, Katherine L. Dixon-Gordon and Matthew T. Tull. "Diagnosis and Characterization of DSM-5 Nonsuicidal Self-Injury Disorder Using the Clinician-Administered Nonsuicidal Self-Injury Disorder Index". *Assessment* 22, no. 5 (2015): 527–39. https://doi.org/10.1177/1073191114565878.

intrusive thoughts'. These reasons can be credited even when the persons injuring themselves are unclear about why they might be doing so.

Such approaches to self-injury contrast starkly and seemingly incommensurably with those of the early modern period. Early modern people saw self-wounding as potentially useful and legitimate for a variety of personal and interpersonal purposes, and rarely a result of mental illness. These purposes—emotional self-regulation, for instance—sometimes matched those listed above, but crucially, reports of self-wounding in the early modern period were almost always of isolated events. People hurt themselves on a one-off basis for a specific purpose, and their actions could not therefore be considered as either a standalone syndrome, or a symptom of another ongoing disorder. Conversely, twentieth-century discourses of self-harm emphasise the repetition of the act over time. The DSM-5 requires that self-injury be undertaken at least five times in a year to be considered a disorder in its own right, though it may occur much less frequently as part of borderline personality disorder, depression or other complaints.

However, despite the dominance of the medical model in approaching this phenomenon, there is also a shift in some circles towards considering self-injury in broader terms as a product of personal, cultural and social factors. Peter Steggals, for example, describes how, as a psychologist by training, he became increasingly interested in the 'ontological' factors which allowed self-harm to fulfil certain functions for individuals:

> I came to think about self-harm as something that emerges out of this 'communicative commotion of everyday life,' as a kind of knot of meaning and action in the 'webs of significance' that surround and pervade us. And as such a knot I began to think about it as something that cannot be reduced to a static and substantive 'thing' at work within us lie a virus or a short circuit in the brain, but rather as a living and changing product of the complex facilitating background of meanings and social relations that culturally embeds and contextualises it.[18]

Rather than any biological facticity, he attributes to self-harm 'a meaningful package of semantics and pragmatics, or what cultural sociology likes

[18] Peter Steggals, "Making Sense of Self-Harm: Exploring the Cultural Meaning and Social Context of Non-Suicidal Self-Injury". PhD diss., Newcastle University, 2013. http://theses.ncl.ac.uk/jspui/handle/10443/2143.

to call a "symbolic action"'.[19] This goes some way to explaining why, for instance, occasional cutting is considered self-harm while a lifelong smoking habit is not, despite the greater physical damage often caused by the latter activity. Self-harm is an action out of place, requiring an analysis that considers the multiple axes (socio-cultural factors, individual circumstances, somatic and psychological responses) which intersect in any experience of self-injury.[20]

Steggals's work shares much with that of Baptiste Brossard, a sociologist who, like Steggals, views self-injury as both a means of self-regulation and a social practice. Through interviews with people who self-injure in various ways, Brossard comes to see this phenomenon less as an affliction from which one suffers, and more as a means of engaging with the world:

> We thus realize that the concerned individuals are not totally acted on by their disorders, as if they were puppets subjected to their pathologies and to their youth as the depiction of self-injury as a syndrome suggests. Within their range of possibilities, they seek instead to adjust their actions to situations that they perceive as problematic. They prove to be 'reasonable.' Although there are alternatives to hurting themselves, their use of self-injury nevertheless denotes a certain social discernment and exercise, and a certain practical sense in setting up their behavior.[21]

People who self-injure do so in part as a result of what Brossard calls 'malaise', and certain factors are likely to contribute to this state of being. Particularly prominent in Brossard's interviews are the role of family conflict and experiences of sexual assault. Emotional communities (the so-called emo or goth communities) are less likely to contribute to self-injury than tabloid reportage might suggest, as Brossard finds that online forums tend more towards encouraging others to minimise or desist from self-injury rather than valorising it. He also views self-injury as 'a practice of social positioning', which is profoundly relational even though it is usually undertaken in secret. This practice, he finds, offers certain advantages over

[19] Steggals, "Making Sense" 16.

[20] Here Steggals draws on Susan Bordo, who uses the term 'axes of continuity' to describe the sociocultural factors which enable the advent of (for example) anorexia nervosa as a psychopathology (Susan Bordo, *Unbearable Weight: Feminism, Western Culture, and the Body* (University of California Press, 1993), 142).

[21] Baptiste Brossard, *Why Do We Hurt Ourselves?: Understanding Self-Harm in Social Life* (Indiana University Press, 2018), 83.

others because it is both transgressive and potentially discreet. Self-injury frequently took place when individuals found themselves unable to acquiesce to an oppressive figure in their lives, yet also unable to retaliate outright.

Self-injury as social practice, as a symbolic action or means of communication—some of the insights generated by *sociological* rather than medical studies of modern self-injury clearly bear striking resemblances to those yielded in this literary-historical analysis. This draws us once again to the question of anachronism, of which the potential hazard is neatly identified by Steggals:

> in apparently discovering a repeating form like self-harm, seemingly present in both contemporary and Biblical psychiatry, we quickly move … to assume the presence of something more natural than cultural and something that while subject to historical variation always remains the same in essence … And of course more often than not this is how self-harm has been represented, not as a meaningful pattern of content but as a mechanical expression of form determined by an underlying realm of biological and psychological facts.[22]

Does this mean that we ought to stop making comparisons between the modern and the historical? Only if we fail to understand its 'content' and instead see it as an urge which simply emerges in different ways at different times: 'the sociocultural icing … secondary to the bio-genetic cake'.[23] While Steggals correctly identifies that 'we are culturally ill equipped to talk about practices of personal distress in general without engaging in medical discourse of some kind', this is a tendency that may be remedied rather than exacerbated by looking at historical self-wounding, if done with care.[24] Early modern people had a radically different concept of both affect and the body. Emotions were not automatically overlaid onto a base of medical discourse as Steggals finds of modern discussions. To use his metaphor, they were the cake, not the icing. As I described in Chaps. 2 and 4, discussions of rhetoric show just how far early modern people understood emotions as not only culturally mediated but potentially generated by art and culture too. The obverse of this belief was the purposeful

[22] Steggals, "Making Sense", 16. My emphasis.
[23] Steggals, "Making Sense", 23.
[24] Steggals, "Making Sense", 26.

adoption of culturally significant behaviours (violence in lament, for instance) in order to both inculcate and express emotions.

We must remember that most of the examples in this book are either explicitly fictional, or moulded to fit a certain kind of narrative. This means that unlike Brossard and Steggals, we cannot surmise that early modern self-injury occurs in response to certain catalysts such as sexual abuse or family tensions. Stories of self-wounding in the sixteenth and seventeenth centuries are also vanishingly rare, and it is impossible to say whether this is because it occurred less, or was deemed less worthy of note. Nonetheless, there is something to be said for the bare recognition that early modern people apparently understood self-wounding as one means by which people sought to change their situations and bridge the gaps between life as it ought to be and life in all its messy reality. In this sense it was not unreasoning but, as Brossard says, 'reasonable'. I do not wish to suggest that the desire to hurt oneself is a universal constant which simply emerges in different guises at different historical moments. However, the ability to hurt oneself is one shared by subjects over time, and for some of those subjects, that ability has proved useful. The uses to which it is turned may be radically different, but the *usefulness* itself is a thread linking (some) early modern narratives about self-wounding to (some) modern narratives about self-harm. The rest of this chapter explores what happens when this thread becomes the basis on which new co-creative interventions are built.

8.3 Self-Injury 'in public'

From 2021 to 2024, I collaborated with partners including Reading University's ArtLab (headed by Professor Tina O'Connell), the charity Self Injury Support and Acta Community Theatre to devise and carry out a number of creative workshops involving people with experience of self-injury. These included a component of looking at historical self-injury, in a spirit similar to that I describe above; self-injury had many frames, of which mental illness was one, but others might be useful in other times and contexts. The aim was not to familiarise the past, but to defamiliarise the present, and so create a space within which discussion could take place.

The purpose of these workshops was initially as a public engagement exercise. By re-historicising self-injury, I hoped, we might be able to emphasise to clinicians the social conditions within which it existed, and which might need to change in order for people to stop using self-injury. For people who experienced self-injury, I considered that knowing the antiquity of the practice, and considering changes in what people had historically considered either injurious or normal, would reduce feelings of shame and isolation. Hilary Powell, whose research addresses the similarities between modern and medieval experiences of voice-hearing, similarly speculates that 'For some voice-hearers, learning that seemingly similar experiences were valorized and celebrated in the past may prove enriching and allow them to re-evaluate their relationship with their voices and their identity as voice-hearers. For others, this revelation may have little worth; it won't change their daily lives or transform their treatment'.[25] In the event, the projects would prove illuminating in more unexpected ways, inviting new connections between the historical and present-day, and challenging me to think about the aims and ethics of such interventions.

The first project began in 2021, after a lengthy period of preparation periodically stymied by the Covid-19 pandemic. It was to take the form of a series of five art workshops run in collaboration with ArtLab, the University of Reading's art outreach organisation. In each of the weekly sessions, I would deliver a short presentation on the history of thinking about self-injury. In response, participants would work in small groups to first create a sculpture from clay, then scan and digitally animate that sculpture, create a soundtrack, and finally present a short video, a piece of digital artwork. Our hope was that by utilising unfamiliar tools such as 3D scanning and animation, the art-making process would be democratised, and there would be no 'best' and 'worst' productions.

This plan, which seemed simple, turned out to be extremely complex to execute. The foremost difficulty was safeguarding. I sought advice from the University's Ethics Committee and from Professor Tim Kent, consultant psychiatrist for the Tavistock and Portman NHS Trust, and followed his recommendations: a small group of maximum fifteen participants; a mental health professional present at all times; a safe time-out space; careful selection of participants; follow-up contact with directions to psychological services if necessary. We also elected to run a 'beta' workshop for various counsellors, academics and other interested parties, in which we could test the processes involved and garner further advice on how best to run the project. This proved valuable, for while the group were engaged by the historical material they were shown—including Portia's

[25] Hilary Powell, "Voices in Psychosis: A Medieval Perspective", in *Voices in Psychosis: Interdisciplinary Perspectives*, ed. Angela Woods, Ben Alderson-Day, and Charles Fernyhough (Oxford University Press, 2022), 98.

self-stabbing and Hieronimo's autoglossotomy—they felt that it was potentially triggering. They also noted the proximity of sharp implements which could be easily secreted into a pocket. These were sensible concerns which we were careful to address before commencing work with potentially vulnerable people. Nonetheless, they also highlighted the level of anxiety which the prospect of self-injury raised in academics and clinicians—an anxiety which would stand in marked contrast to the attitudes of our later workshop participants. Perhaps unsurprisingly, these professionals felt a strong moral responsibility to prevent self-injury as well as to understand it.

With the lessons of the beta session under our belts, I set about selecting participants for the forthcoming sessions. The recruitment process had not been straightforward. At the outset of the project, I naively imagined that mental health organisations would direct their service users to us, as offering an artistic experience which was beyond the scope of their funding to provide. Quickly, however, we realised that these overstretched organisations were not inclined to refer vulnerable people into a process which had not been tested, despite reassurances of the consultation that had taken place in the sessions' formulation, and the professional advice which would be available. A new tactic was needed; we opened up the call to include not only those with direct lived experience of self-injury, but anyone with an interest in the topic, hoping to attract friends and family members of people who self-injured as well as interested professionals such as teachers and social workers. Gradually, responses began to increase until it became clear that we had gathered far more interest than we could accommodate if the sessions were to run safely. It was also evident that the vast majority of that interest came from young women between the ages of eighteen and twenty-five. Using a questionnaire, I unscientifically selected participants with an eye to variety: different ages, backgrounds, genders and reasons for wanting to join the project. Informed by the beta session, I called each prospective participant to explain the aims of the project, outline the shape of the sessions, and, as far as possible, gauge whether any mental health difficulties were likely to render them at serious risk from a process which would necessarily involve difficult subject matter and challenging conversations.

Over five sessions, we looked at aspects of my research more or less related to the topic of self-injury: the four humours; ethnography and bodily alteration; Renaissance beliefs about gender and sexuality; forms of protest. The participants responded enthusiastically to these talks, and were notably eager to construct a narrative which spanned more than the early modern period, to understand why and how attitudes had changed to become, as they perceived, more stigmatising and far more fearful. By contrast, the art workshops became a space for conversation and play, often full of chatter and laughter despite the subject matter. We attempted to measure participants' attitudes to NSSI over time (an attempt which failed for reasons I describe below), and recorded parts of the workshop discussions. What we did not measure, however, was the positive effect of simply providing a space in which people could come together to make art once a week, after a year of social distancing.

Many lessons were learned during this process, not least of which was the difficulty of retaining participants who had entered an unusual situation largely on the strength of curiosity and hope of some therapeutic benefit. Recruiting for public engagement is notoriously difficult, and I would usually oversubscribe a session by around 30 per cent to ensure an adequate turnout. That was clearly untenable in this case, and after the packed first session, attendance dipped to between 30 and 70 per cent of capacity for any remaining sessions. A number of factors affected this pattern. We were at the height of the Omicron variant of Covid-19, and given that many of our participants were either public-facing workers or students, several had typically contracted Covid or other viruses in any given week. Despite our attempts to explain how the sessions would pan out, there were inevitably those who attended and found that the process was not what they had hoped. In addition to these universal problems, however, the particular subject matter was one in which people were bound to encounter difficult feelings, and to withdraw as a result. The stigma around self-injury had most likely combined with the rigorous process of screening which had taken place before the first workshop to make these people feel that they could not express their reasons for not wishing to return. Those who skipped sessions and did not give reasons of illness simply ceased communication, and we were bound to respect their desire to avoid any perceived confrontation. As a result, our attempts to compare participants' wellbeing and attitudes at the start and end of the process were rendered meaningless. After five weeks, we were comparing few of the same people. Our conclusion that the workshops had allowed participants to talk about NSSI more openly was based on observation and discussion rather than provable change.

* * *

This first phase of my publicly engaged research appeared successful, and won a University prize. Yet the broader project of interrogating the links between early modern and modern self-injury felt unfinished. The next steps were unclear, until an email arrived from a trustee from the charity Self Injury Support (SIS). They had seen the publicity around the first project and the award, and suggested I meet with a member of their team to discuss possible collaborations. SIS had independently been engaged in historicising self-injury, particularly thinking about the history of their own organisation. They were also actively seeking to provide more creative opportunities for their clients. Over several months, we devised a new project. With the community theatre charity Acta Bristol, we would run a six-month series of weekly workshops involving people with lived experience of self-injury. In the course of the workshops, we would reflect on the links and disjunctures between perceptions of self-injury in the past and today, and work towards scripting, staging and performing a theatre piece.

The advantages of working with partners who were specialists in therapeutic arts and safeguarding were obvious. SIS now oversaw the safety of the project, screening applicants and providing two support workers to be present at (and participate in) each session. Acta recommended a freelance theatre facilitator to run the sessions, and I attended once a month to provide historical context and prompts for the next month's discussions, focussing in particular on a timeline of attitudes towards self-injury and on particular literary examples such as *Tamburlaine* and *Julius Caesar*. Once again, however, the reality was less straightforward than we had envisaged. Recruitment was painfully slow, and several participants joined only to leave again for various reasons. Two months into the project, it was clear that it was not working as envisaged.

At this point, the involvement of a reflective practitioner proved incredibly helpful. This is routine practice for many charities, particularly those working in mental health, but remains uncommon among academics engaging in similarly challenging projects. Through discussion with the practitioner, we reassessed what we were hoping to gain from the project. We wanted participants to join because *we* felt it was a valuable opportunity, but the lack of sign-ups indicated that what we had planned was not what potential participants found useful or feasible.[26] To continue attempting to recruit in the same way would be serving our own purposes—particularly the desire for 'impact'—rather than authentically responding to the needs of those with whom we sought to engage. This realisation led us to reconsider the power balance which we had embedded in our plans. What if, instead of 'engaging with' participants from a stance of authority, we instead engaged in genuine co-creation with our tiny cohort of articulate, creative and highly committed participants, without waiting for the 'real work' to begin? Broader public engagement might come from the resultant outputs, but the participants' role was now to help communicate their knowledge and experience, not to be impacted upon.

Such an approach transformed the dynamic of the project. We began to consider how we could facilitate co-creation at different levels of involvement, and opened an online call for audio, written and artistic responses to the questions we had been addressing as a group. We received submissions from people unable to participate in the workshops for reasons of geography, health and logistics; many would be incorporated into the final work. We also reconceived what a theatre piece might involve; playing to participants' strengths, we now included puppetry, music and spoken word contributions, alongside a small art exhibition. These changes required almost no logistical change to the programme, but a readjustment of our expectations and ideas of success. In the history-focussed workshops we discussed Portia and Tamburlaine, the humours, the evolution of asylums, medieval saints and self-flagellation. But it was Portia who became emblematic of the project; a character whose self-injury was explicitly and unapologetically attention-seeking. Why, we asked, did Portia need to hurt herself to be heard? Was her self-injury a 'feminine' tactic? At what point had accusations of 'attention-seeking' become a way to delegitimate a person's actions rather than an understandable and even laudable response to emotional distress? The group devised a script in which five fictional characters, including Portia and Tamburlaine, gathered in a support group to take turns in telling their stories, interspersed with discussions which were largely improvised. The second half of the

[26] The question of payment for participants is a live debate. While I generally pay participants for their expertise, it was felt that this would be inappropriate in this project for a number of reasons (again, I benefitted from the expertise of SIS in this case). To mitigate this effect, we paid participants' costs, including providing taxis, and supplied meals during rehearsals and performances.

performance was dedicated to online submissions; poems, written pieces and spoken word.

8.4 The Impact of Impact

These projects, particularly the latter, undoubtedly touched those involved. Among the participants of the theatre project, all but one reported feeling more able to talk to either friends, family or medical professionals about their experiences with self-injury. When asked what, if anything, they would change about the process, the most common response was that they wished it could continue indefinitely.[27] Change wrought in the audience of the performances was harder to measure, but in written qualitative feedback they reported being surprised that self-injury *had* a history, and 'moved', 'inspired' and 'soothed' by seeing it talked about openly. Several commented that they now planned to discuss the topic with friends and family, including one person whose child had experienced self-harm, who stated that they now wanted to apologise to them for their lack of understanding. At the time of writing, the project continues to have an afterlife through an exhibition in Glenside mental health museum, Bristol, and through a related group run by Self Injury Support. It also provides a model for further partnerships, including with Oxfordshire Recovery College.

This work was, therefore, 'impactful', in the sense of 'impact' provided by the United Kingdom's Research Excellence Framework (REF): 'an effect on, change or benefit to the economy, society, culture, public policy or services, health, the environment or quality of life, beyond academia'. Yet, I argue that 'impact' may be an inadequate metric for measuring the benefit of such projects. To have an effect *on*, make change *to* or impart a benefit *to* audiences beyond academia assumes that it is only those outside the academe who require amendment. This point has not escaped the notice of scholars. Jennifer Sandlin, Jake Burdick and Emma Rich argue that public engagement based on the logics of pedagogy 'do not necessarily lead to education in the service of publicness', because they position the educator in a position of superior knowledge, and therefore curtail the

[27] Abrupt endings to projects are a recognised problem with public engagement, especially with vulnerable groups. This project had a debrief session and participants were signposted to further services—including other creative groups—run by Self Injury Support.

directions in which learning might be allowed to go.[28] Most public engagement, they contend, is really public pedagogy, and 'Most conceptualizations and enactments of public pedagogy aim to effect a certain intended outcome for or disposition within the individual'.[29] The aims of public engagement too often presuppose a certain way in which participants will respond or collaborate, thus curtailing genuine curiosity and collaboration. Furthermore, 'Under this pedagogical umbrella, learning is always future-oriented, as learners are urged to *become* something, and to embrace an ideal of being or doing *something better* in the future'.[30] That urge is, they argue, tied to neoliberalist capitalism, the idea of improving oneself to become a more productive part of the workforce.

If this model is unsatisfactory, what can be put in its place? What might it look like if 'impact' operated in both directions; if academics were impact*ed* as much as impact*ing*? Such an approach would undoubtedly pose challenges for the planning and delivery of engagement projects, yet an increasing number of academics and practitioners have argued for such a rethinking. Sandlin et al., for instance, propose new 'public interventions' based on tenets including 'not attempting to facilitate a particular outcome of learning, but rather enacting the role of interrupter [of prevailing discourses]'.[31] Importantly, they suggest that

> scholars might learn to sit uncomfortably in the space that lies between apprehending and eclipsing those pedagogical interventions and pedagogues, especially if we want these sites to retain their qualities 'of human existence' that make 'new histories and emergent identities possible'. If we cannot or will not remain in these liminal spaces, scholarly work on public pedagogy will likely become enervated of its possibility, its alterity to commonsensical forms of education, and its very viability as a political site of resistance.[32]

[28] Jennifer A. Sandlin, Jake Burdick, and Emma Rich, "Problematizing Public Engagement within Public Pedagogy Research and Practice", *Discourse: Studies in the Cultural Politics of Education* 38, no. 6 (2017): 823–35, (830), https://doi.org/10.1080/01596306.2016.1196343.

[29] Sandlin, Burdick, and Rich, "Problematizing Public Engagement", 830.

[30] Sandlin, Burdick, and Rich, "Problematizing Public Engagement", 831.

[31] Sandlin, Burdick, and Rich, "Problematizing Public Engagement", 833; see also Isabelle Heyerick, "Is There an I in Impact? Considering the Two-Way Process of Public Engagement", *Exchanges: The Interdisciplinary Research Journal* 7, no. 1 (2019): 82–93, https://doi.org/10.31273/eirj.v7i1.520.

[32] Sandlin, Burdick, and Rich, "Problematizing Public Engagement", 833.

Forms of engagement which utilise the creative arts are particularly apt to these aims. Co-creativity, as Tischler states, is 'characterized by shared decision-making and ownership, reciprocity and relationality'.[33] It recognises non-academic participants as experts by lived experience, and the academic as potentially equally impacted by engagement, though often in unmeasurable ways. This immeasurability renders co-creative, bidirectional impact difficult to justify in the terms seemingly encouraged by current frameworks, leading some to warn that impact measurement 'carries the danger of squeezing out participatory and collaborative research at a local scale with small organisations, grassroots movements and activists'.[34] Co-creation is also intrinsically unpredictable. As I learned at first hand, to be genuinely responsive to the needs and desires of co-creators means admitting the possibility that the impacts of a project may be very different to those initially envisaged. To insist on *impacting* non-academic participants in a particular way risks not only slipping into pedagogy, but instrumentalising both the process and those involved therein. Yet there is also a desire to see 'value for money' in such projects which limits freedom to act responsively to the insights of participatory research. The complete solution is far from clear. However, it might involve: (1) co-design of participatory research from the very earliest stages, involving *all* the potential interested groups; (2) willingness by funders to accept the real possibility that a project may not deliver in the way intended; and (3) readiness among academics to allow that even the most meticulously planned work may veer off course, and to accept these 'failures' as informative.

8.5 Healing the Breach

This is a long way from Shakespeare. There is much that cannot be accomplished through 'unhistoricism', or through the dialogue with modern phenomena that it encourages. Portia's authenticity can only be fully understood with reference to the persuasive tactics of classical rhetoricians, while to appreciate Hieronimo's tongue-biting requires knowledge of Elizabethan neo-Stoicism. Furthermore, the internal world of people in

[33] Victoria Tischler, 'Health Humanities and the Creative Disciplines', in *The Routledge Companion to Health Humanities*, ed. Paul Crawford, Brian Brown, and Andrea Charise (Routledge, 2020), 88.

[34] Rachel Pain, "Impact: Striking a Blow or Walking Together?", *ACME: An International Journal for Critical Geographies* 13, no. 1 (2014): 20, https://doi.org/10.14288/acme.v13i1.986.

the past is mediated by culturally specific ways of emoting and thinking that are now foreign to us, as in tales of lament. Conversely, people who experience self-harm in the twenty-first century may perceive it as anything but 'useful'. It is important to realise that the participants in public projects on this topic are self-selecting; many people exist for whom self-injury feels far more medical than social, and for whom early modern representations of this topic seem very far removed from reality. Nonetheless, I hope to have shown that there exists some common ground between these two poles. Modern and early modern share—at least sometimes—a sense that self-injury seeks to bridge the gap between one person's experience and another's by translating ineffable emotional experiences into tangible physical ones. This is, of course, a receding target. The exact nature of one person's physical pain cannot be known by another, any more than their mental state. But self-injury brings attention to the universality of pain *as* a human experience, and seeks connection on that basis. It is attention seeking, but this is not to say that it is inauthentic. Rather, for people in early modern England, self-wounding simply served to say 'look, *here*'.

Appendix

I Am Worthy, Play Devised by Participants of the 'Self-Injury Through Time' Theatre Group, 2024

Facilitator sets up stage as the audience settles. Brings out chairs one by one from backstage, places them in a semi-circle and sets out tea and coffee trolley.

As participants arrive from stage, they quietly greet them and introduce them to the space.

Characters are all seated.

OLLIE:

I think that's everyone for today. Shall we make a start?

First off, thank you all for coming today, I hope we can make this a nice evening of talking and hearing each other's stories.

Before we get lost in conversation, I just wanted to go over the group agreement on how we will make this space as safe as possible. Everything discussed here tonight stays in this room to keep confidentiality.

As you can see, we've all come from different backgrounds and experiences.

So I'd like us to keep respectful when listening to each other.

A. Skuse, *Hurt Feelings*, Palgrave Studies in Literature, Science and Medicine, https://doi.org/10.1007/978-3-032-04565-2

I thought we could start by going round and say your name, pronouns—if you'd like, what you do and which animal describes you best and why.

OLLIE: I'll go first. Hi, I'm Ollie. I'm non-binary and use they/them pronouns. If that's new to you, feel free to ask me any questions. I also have lived experience with Self-harm/injury. I'm most like a fox coz I'm always up at night, rarely seen in public and quietly confident. Who would like to go next?

Shall we start over here? **signals to Portia.* *

PORTIA: say your name, pronouns—if you'd like, what you do and which animal describes you best and why.

CARLOS: say your name, pronouns—if you'd like, what you do and which animal describes you best and why.

TAMBURLAINE: say your name, pronouns—if you'd like, what you do and which animal describes you best and why.

BEN: say your name, pronouns—if you'd like, what you do and which animal describes you best and why.

HELENA: say your name, pronouns—if you'd like, what you do and which animal describes you best and why.

OLLIE: Awesome, some great animals in the space today and I really feel like I got a sense of who you all are so thank you and hopefully we can have some interesting discussions today.

I'd like us to start with a question…

Why do we think people self-injure?

Everyone pauses and looks around uncertainly.

TAMBURLAINE: For me, it made me feel more powerful.

CARLOS: Powerful? It doesn't make me feel powerful. Sometimes, the only way I can distract myself from my mental pain is to create physical pain on myself. I can understand it more if I can see it this way.

BEN: I know what you're saying, I've never seen it as powerful before, I see it as vulnerability.

TAMBURLAINE: For me, there was an element of performance. In my time, we had to make a statement to be listened to.

PORTIA: Me too. I did it to show my strength and what I could endure. It made people stop doubting me.

HELENA: For me it was about what I *couldn't* endure. When I couldn't endure what was in my head, I used self-harm to process it. It was about not feeling good enough.

OLLIE: And why do you think that is?

HELENA: I think it comes from social media.

TAMBURLAINE & PORTIA: **confused* Social media?*

BEN: It's like socialising but online.

TAMBURLAINE: That sounds like a terrible idea. Why would you want to socialise with people without seeing them? How should you know what they really mean?

HELENA: And it's all instant! Instant access to each other all the time-it's exhausting.

CARLOS: I see idiots posting online all the time. People post about their self-harm. What's that all about? It's like they're asking for attention.

HELENA: I don't see people posting their self-harm. But when I see people moving in together, getting new jobs, celebrating all the positives in their life. It's like I have to constantly compare myself and I feel left behind.

OLLIE: I think it's important to remember that social media often only shows the best of people's lives.

BEN: Yes, we only see what people want us to see.

PORTIA: I'm sorry, I still don't understand.

TAMBURLAINE: Me neither. Why would you want to see what other people are doing ALL the time? Is there no ambiguity left in this world?

CARLOS: *rolls his eyes* No wonder people are unhappy!

OLLIE: This is an interesting thought. So, do we think self-injury is a modern-day issue because of social media?

HELENA: Yes!

TAMBURLAINE: No!

BEN: I think it's perceived as a modern-day issue. Now it is talked about and given a label but maybe in the past it existed, but we never heard about that?

HELENA: I studied history at Uni and never once did I hear about self-injury then.

PORTIA: I think you need better history teachers!

OLLIE: Portia, it sounds like you have more to share. Would you like to start us off by telling us more about why you're here?

PORTIA: I'd be pleased to.

As Portia starts her story, all characters get up and move set.

Portia walks downstage as she starts speaking Carlos and Ben go to the band area.

My story is a story of great dedication and honour. My self-inflicted pain is not something I'm ashamed of, for it brought me what I desired. In the first century BC, women are seen as nothing but passive and submissive wives. But I was different. I was the great daughter of Cato; a woman Brutus was proud to take as a wife. Though he did not seem to trust in me.

Music starts.

We were supposed to be a team—I thought husband and wife were supposed to share everything together? But he was talking in secret, hiding things from me, he wouldn't let me in. I felt so unimportant, looked down on. Even though we were married, that didn't mean he would confide in me. I was supposed to just know my place.

But I was smarter than they believed, smarter than my husband believed. You see, I understood the political climate and my husband's role in it. Increasingly, I became aware of his involvement in the conspiracy against Caesar. I also saw the doubts he was having about it. But, still, he would not confide in me. Still, he would see me as just a woman, someone to just be quiet and submit to him. So, I spoke in the only language these brave men seemed to understand. I took a sharp knife to my thigh and let that do the talking.

I bore this pain quietly but shared the wound proudly. In this act, I showed my brave determination, my unwavering devotion to my husband and his cause. I proved my worthiness and my strength in this act of self-injury. He couldn't deny that anymore. Actions speak louder than words. My words of encouragement and support could never truly get through to him. My action to inflict physical pain on myself told him everything he needed to know about me.

My identity wasn't enough. The fact that I was born from such a strong father wasn't enough to prove my worthiness, to make me more than just a woman. My wound finally made him see me in a different light, it finally got his attention. He could finally see what I was capable of enduring. In modern times, it seems that attention seeking is viewed as a bad thing. But I would have been left silently in the dark had I not fought for the recognition I deserved.

Despite my own fears and concerns, despite the danger I was put in, my support for my husband always remained strong. I'm glad I harmed myself, it meant that Brutus could never doubt me again. He could never doubt my loyalty. He knew that I could be trusted. He never saw the self-harm as a bad thing or something to be concerned about. The fact that I willingly endured pain for him proved my commitment to him and his cause. If I had to, I'd do it all again. My self-harm showed my strength, not my weakness.

Music subsides and characters return to their chairs

OLLIE: Portia, thank you for sharing with us.

HELENA: I think that was really brave of you. Really strong.

TAMBURLAINE: Strong? I didn't know women could be "strong". In my day, strength came from men. From soldiers.

CARLOS: So, you did all that just for your husbands' attention?

PORTIA: No, I did it to prove myself. To save myself and my husband, I don't see shame in that.

TAMBURLAINE: Did he?

PORTIA: Did he what?

TAMBURLAINE: Did *he* see shame in that?

PORTIA: No, my husband never shamed me for my self-harm, he saw it as a noble act. But ... maybe shame led me to it because I didn't feel good enough for him. No matter what I did, I couldn't prove myself.

BEN: I think you were noble. I respect how you can show up for yourself. You shouldn't feel ashamed.

OLLIE: Shame an interesting word. Why do we think some people feel they have to hide their self-injury? Carlos?

CARLOS: Some self-harm attempts may be conscious, some might not. Sometimes, there's something or someone inside you that you want to rip out, so much that you'd be prepared to end your own life in order to be free of it. Other times it's impulsive and you end up in a pool of blood before you've even had the chance to think of the consequences.

OLLIE: How about you Helena?

HELENA: Well I told no one, because I held no sense for what I'd done and certainly no voice. Now I understand that self-harm was the only "voice" I had.

TAMBURLAINE: "Actions speak louder than words."

HELENA: Not always. But I think shame comes from the fact people don't understand it. It goes against our natural instincts to look after ourselves. I don't think people realise how much self-injury buries you and sinks you in to a secret life.

BEN: Exactly. Instead of drawing attention to the emotional pain and hurt I was in, self-injury just wrapped me in more darkness and the shame kept me there.

OLLIE: Well, whilst we are on the topic of shame, this might be a good time to share more about me. I'd like to tell you a story.

As OLLIE speaks, the characters disperse. CARLOS and BEN to the band, and OLLIE walks to upstage right raised platform, rest of set up flats for shadow puppetry. HELENA and TAMBURLAINE to OHP (overhead projector).

OLLIE: For me, I've felt like I needed to hide parts of myself in the past. I felt that I wasn't worthy of having challenges or struggles or people's time, so of course, the answer was to pretend I was ok.

Lights up on OLLIE upstage right and band upstage left, OHP on.

OLLIE: Once upon a time there lived a Minotaur. They lived deep inside a cave. The cave had a tiny opening to the outside world where there were birds, flowers, grass meadows and sunlight. Inside the cave, it was full of dark corners, sharp edges and twists and turns. The Minotaur lived deep inside, right in the centre. They lived alone, in the darkness, but acutely aware of the opening in the cave.

The Minotaur worked very hard to either stay as quiet as they possibly could so that nobody would ever know that they were there. Or they made themselves as frightening and terrifying and vicious as they could be. The aim was the same—to keep anyone and everyone far away from them because, it was way too scary for the Minotaur to consider letting anyone in.

Whenever someone got close enough to the entrance of the cave, the Minotaur would spit out arrows with deadly precision. Until, one day, they spat out an arrow and heard the most sorrowful, sad cry. The Minotaur had never heard a sound like it and it stirred them from the centre of the cave. They reached the opening and found a young girl, lying on the floor, injured from the arrow wound. She was crying out for help. When she saw the Minotaur appears from inside the cave, her eyes held a hope that the Minotaur could help. A belief that the Minotaur had the goodness in their heart to help, not to hurt her further or run away.

In that hopeful moment, the Minotaur began to transform. Their two bull-like horns turned into one Unicorn horn and their hooves planted to the ground as they now stood on all fours, as a black unicorn with a golden horn.

The Unicorn instinctively knew what to do and where to go. They picked up the girl and galloped to her village where they were met with care and concern. The girl's people nursed her and her wounds and fed the Unicorn, tending to them both gently. The girl's people were so grateful for the Unicorn saving her. The Unicorn felt this tenderness and care.

But inside, they still felt the shame of being the Minotaur. Of hiding themselves and what they really were.

Soon, the girl healed and began to move around the village with her new Unicorn friend. But one day, the villagers gathered around them. "We know what you really are", they told the Unicorn. The Unicorn felt a sense of horror, shame and panic rising. The Villagers told the Unicorn that they knew who they were—they were the Minotaur. Suddenly, the Unicorn's horn turned back into two bull-like horns, they stood back up onto two feet. The Minotaur went to run, to run and hide back in their cave where they felt they belonged.

"WAIT!" a little girl's voice said. "Where are you going? We don't want to you leave. You saved me." The Villagers all nodded in agreement. "We know you have a good heart. You don't need to be afraid anymore, we want you to stay. To be part of our Village".

Music stops and cast resets the stage and returns to their seats.

Through hiding, I lost me… or maybe, I found me once I started living life FOR me.

OLLIE: Has anyone had experience with hiding parts of themselves and then revealing it?

PORTIA: I never thought it was something I had to hide, I made very sure that my husband saw it and understood why I did it.

BEN: My hiding comes from a fear of what other people will think.

OLLIE: And why do we think that is?

CARLOS: Because there's still so much stigma around it, obviously. My shame comes from being at school and getting bullied. Having to wear long sleeved t-shirts and jumpers to hide my arms.

HELENA: Because people aren't willing to bring up the topic and just be direct. We shouldn't be afraid of having difficult conversations.

PORTIA: That's not always been the case.

BEN: But back in your days, life was more violent, wasn't it? Weren't people always killing each other?

TAMBURLAINE: I suppose it was. And there wasn't all this terminology back then. We fought to survive, and we spoke openly about it.

PORTIA: This is true. There was no assumption that someone was mentally ill because they harmed themselves. If anything, it was a physical problem—like your humours being imbalanced.

HELENA: Your humours?

PORTIA: Have you never heard of the four humours?

ALL: "No", "I haven't", "what are they?"

PORTIA: The 4 humours were used in medicine in our time. The humours were carried in your blood stream and depending on which was the most present, you'd feel a certain way and have the physical symptoms to show.

TAMBURLAINE: Where do you think the term "sense of humour" comes from?

PORTIA: Exactly! Playwrights often used them to inspire characters—like in Shakespeare's plays.

TAMBURLAINE: Or Christopher Marlowe...

PORTIA: Probably!

OLLIE: Tamburlaine, it sounds like you know more about this. Would you like to share your story?

TAMBURLAINE: Gladly.

Characters stand and begin to move set. Music starts from band.

OLLIE, HELENA and PORTIA put on cloaks to assume roles of CALYPHAS, CELEBINUS and AMYRAS.

I must paint a picture of what life was like for me to show why I'm here. You see, I was emperor and conqueror in Persepolis/Persia, what you now call Iran, in 1300. They called me Tamburlaine the Great. I'm an ambitious man. An invincible man. So many battles I bought and not once was I injured.

MONOLOGUE

Which leads me to introduce my sons- Calyphas, Celibinus and Amyras.

Now, my boys, listen to me. I'll teach you to be warriors. To sleep on the ground, march through all weathers, to sustain the scorching heat and freezing cold. I'll teach you how to scale a castle wall, besiege a fort and undermine a town. When this is done, then you are soldiers. And worthy sons of Tamburlaine the Great.

CALYPHAS: But Father, this sounds dangerous. We might be killed or injured.

TAMBURLAINE: Are you my son or not? Are you a coward, for you speak of fearing death or wounds?

Just look at me. I have conquered Kings. I have marched around the Earth, battling various Armies. And here I am, not a single scar. I am clear from any wound and haven't lost a dram of blood. But watch this, this will teach you.

He turns towards the back and imitates cutting his arm.

Blood is the God of War's rich livery. Now, I look like a soldier, and this wound is a symbol of grace and majesty. As I sit here with my diamonds and pearls from wealthy India, and my robe from Africa.

Come boys. Come and touch my wound. Wash your hands with my blood, whilst I sit smiling to behold the sight. My boys, what do you think?

CALYPHAS: I don't know what to think.

TAMBURLAINE: Go on…

CALYPHAS: I think it's a pitiful sight.

CELIBINUS: It's nothing. Give me a wound, Father!

AMYRAS: And me, another!

TAMBURLAINE: Come then, give me your arm.

CELIBINUS: Cut it as bravely as you did your own. *He holds out his own arm.*

TAMBURLAINE: My boy. You shall not lose a drop of blood before we meet the next army. My speech of war and this wound that you see, was to teach you boys to be courageous. To follow the great Tamburlaine.

Music ends.

As TAMBURLAINE continues speaking, characters return to their seats.

And that is how I gained this wound on my arm. It was a lesson to my sons. So that they understand the bravery and strength it takes to be a soldier.

Group Discusses

OLLIE: Thank you Tamburlaine- it's interesting to hear a different perspective. It seems like not everyone feels the same sense of shame that others of us do or have done. Would anyone like to add anything?

PORTIA: It seems to me that people are too afraid to talk about it, to understand what it means for different people and that's where the shame comes from.

HELENA: But at least you knew what you wanted from it.

PORTIA: True. But I only did it for my husband… and, I don't know, maybe that's sad in itself. That I actually had to injure myself in order to get his attention.

BEN: **to TAMBURLAINE** And who did *you* do it for?

TAMBURLAINE: Well, I did it for myself. It made me look stronger. Like I could withstand anything. Any pain, any challenge, and any battle.

CARLOS: But what if no one else sees it as strong? Everyone else sees it as a weakness. Most of those who I reached out to only saw it as attention seeking.

HELENA: Or they just don't know how to help.

OLLIE: So, what can we try and change so that people *do* understand?

TAMBURLAINE: I'd say don't just follow the attitudes you were taught by society, make your own mind up.

CARLOS: We shouldn't be afraid to share our stories or feel like we need to shy away from the difficult conversations.

PORTIA: I think we need to stop seeing it as "attention-seeking". We are all seeking attention and we communicate this in different ways.

HELENA: I think we need more education and more activism from people with lived experience and their allies.

BEN: Like with most sensitive social issues it would need to start in school—a more open dialogue around self-injury in secondary school, no sugar coating.

OLLIE: HELENA*, we've not heard your story yet. Would you like to share something?

HELENA: Ok. I can share. For me, I find writing things down has really helped me. I use this journal and like to write down anything I'm feeling to help to process it. So I'd like to read something that I've written, if that's ok?

Lights go down, apart from focus on HELENA.

Quiet music starts.

This poem is called "Why I'm quitting self-harm"
Because I recognize my arm is not a cutting board
Pieces of Styrofoam
Or a scrapbook of every time I couldn't bear to cope
No, my body is a temple of the all Holy Ghost
And it's mine to care for
Because I know that God made my form intricately
And yet I'll punish myself for things that are fixed permanently
Or just use it to breathe
Because what they don't teach you in your self harm career
Is how often you'll feel like you're suffocating?
Your throat coated with tears
So you'll cut open slits to air yourself out
The blood is a river that makes its way round the bends of the flesh
And now your lungs fill up again—
Until it's a full-blown, honest to heavens actual addiction
I am so used to being a flooded reservoir and a runaway car
That speeds down the street
A liability

But I'm staking a claim on this holy place of worship
Because I know I am worthy.
Because I'm finally choosing me.
MUSIC stops.
Characters return to their seats.

OLLIE: Thank you Helena*. I love that final line "because I know I am worthy". I think sometimes we can question our worth. We forget that everyone is as worthy as each other of being here.

HELENA: It wasn't easy. But soon, I stopped judging myself and started realising that my worth came from inside. I think others here already knew that.

OLLIE: That's great to hear. It seems like many of you have learnt something from each other today. With this in mind, I'd love to hear from you all, what is one thing *you* are worthy of?

PORTIA: I am worthy of being heard
CARLOS: I am worthy of peoples' time
BEN: I am worthy of being seen for who I really am
TAMBURLAINE: I am worthy of being vulnerable
HELENA: I am worthy of taking up space
OLLIE: I am worthy of....

As each character repeats their lines, they stand, pick up their chair and exit the stage.

PORTIA: I am worthy of being heard
CARLOS: I am worthy of peoples' time
BEN: I am worthy of being seen for who I really am
TAMBURLAINE: I am worthy of being vulnerable
HELENA: I am worthy of taking up space
OLLIE: I am worthy of....

CARLOS and BEN exit stage left towards the band area. OLLIE is last on stage and takes a moment to glance back at the empty room before they EXIT.

Bibliography

Abbot, John. 1623. *Jesus Præfigured, or, A Poëme of the Holy Name of Jesus in Five Bookes*. ProQuest, UMI.

Adams, John. 1700. *An Essay Concerning Self-Murther*. Tho[mas] Bennet.

Adler, Patricia A., and Peter Adler. 2011. *The Tender Cut: Inside the Hidden World of Self-Injury*. New York University Press.

Ahmed, Sara. 2014. *Cultural Politics of Emotion*. Edinburgh University Press.

Alexander, Gavin. 2007. Prosopopoeia: The Speaking Figure. In *Renaissance Figures of Speech*, ed. Sylvia Adamson, Gavin Alexander, and Katrin Ettenhuber. Cambridge University Press.

Alexiou, Margaret. 2002. *The Ritual Lament in Greek Tradition*. Rowman & Littlefield Publishers.

Allen, Hannah. 1683. *A Narrative of God's Gracious Dealings with That Choice Christian Mrs. Hannah Allen (afterwards married to Mr. Hatt.). Reciting the Great Advantages the Devil Made of Her Deep Melancholy, and the Triumphant Victories, Rich and Sovereign Graces, God Gave Her over All His Stratagems and Devices*. John Wallis.

Ancillon, Charles. 1740. *Italian Love: Or, Eunuchism Displayed. Describing All the Different Kinds of Eunuchs; Shewing the Esteem They Have Met with in the World, and How They Came to be Made So, Wherein Principally Is Examined, Whether They Are Capable of Marriage, and of They Ought to be Suffered to Enter Into That Holy State … Occasioned by a Young Lady's Falling in Love with One, Who Sung in the Opera at the Hay-Market, and to Whom She Had Like to Have Been Married. Written by a Person of Honour*. Printed for E. Curll.

Anderson, Miranda. 2015. *The Renaissance Extended Mind*. Springer.

Aristotle. 1997. *Aristotle's Poetics: Translated and with a Commentary by George Whalley*. McGill-Queen's University Press.

A. Skuse, *Hurt Feelings*, Palgrave Studies in Literature, Science and Medicine, https://doi.org/10.1007/978-3-032-04565-2

Armitage, David. 2022. In Defense of Presentism. In *History and Human Flourishing*, ed. Darrin M. McMahon. Oxford University Press. https://doi.org/10.1093/oso/9780197625262.003.0003.

Bacon, Francis. 1638. *Historie Naturall and Experimentall, of Life and Death. Or of the Prolongation of Life*. John Haviland for William Lee, and Humphrey Mosley.

Bacon, Francis. 1651. *A True and Historical Relation of the Poysoning of Sir Thomas Overbury*. T.M. & A.C. for John Benson and John Playford.

Baker, Richard, Sir. 1643. *A Chronicle of the Kings of England, from the Time of the Romans Goverment unto the Raigne of Our Soveraigne Lord, King Charles Containing All Passages of State or Church, with All Other Observations Proper for a Chronicle*. Printed for Daniel Frere.

Baldwin, Geoff. 2001. Individual and Self in the Late Renaissance. *The Historical Journal* 44 (2): 341–364. https://www.jstor.org/stable/3133611.

Bale, John. 1551. *The First Two Partes of the Actes or Unchast Examples of the Englysh Votaryes Gathered Out of Their Owne Legends and Chronycles by Johan Bale*.

Barbagli, Marzio. 2015. *Farewell to the World: A History of Suicide*. John Wiley & Sons.

Bargu, Banu. 2014. *Starve and Immolate: The Politics of Human Weapons*. Columbia University Press.

Barksdale, Clement. 1675. *Three Ministers Communicating Their Collections and Notions. Touching Several Texts of Scripture … Wherein the Law and Gospel … in Short, the Substance of Christianity Is Set Forth*. Printed for Samuel Keble.

Barrett, Timothy. 2015. Storying Bourdieu: Fragments Towards a Bourdieusian Approach to 'Life Histories'. *International Journal of Qualitative Methods* 14 (5): https://doi.org/10.1177/1609406915621399.

Barthorpe, Amber, Becky Mars, Paul Moran, and Lizzy Winstone. 2020. Is Social Media Screen Time Really Associated with Poor Adolescent Mental Health? A Time Use Diary Study. *Journal of Affective Disorders* 274:864–870. https://doi.org/10.1016/j.jad.2020.05.106.

Bate, Jonathan. 2018. Enacting Revenge: The Mingled Yarn of Elizabethan Tragedy. In *Doing Kyd: Essays on The Spanish Tragedy*, ed. Nicoleta Cinpoeş. Manchester University Press.

Beecher, D. A. 1988. The Anatomy of Melancholy in Book I of the Faerie Queene. *Renaissance and Reformation* 24 (2): 85–99.

Betancourt, Roland. 2023. "To Clarify an Everchanging Present" – Hyperchronicity, Anachronism and the Relationship Between History and Social Science. *LSE Impact Blog*, May 10. https://blogs.lse.ac.uk/impactofsocialsciences/2023/05/10/to-clarify-an-everchanging-present-hyperchronicity-anachronism-and-the-relationship-between-history-and-social-science/.

Blankaart, Steven. 1684. *A Physical Dictionary*. J.D.

Bordo, Susan. 1993. *Unbearable Weight: Feminism, Western Culture, and the Body*. University of California Press.

Bound Alberti, Fay. 2006a. Emotions in the Early Modern Medical Tradition. In *Medicine, Emotion and Disease, 1700-1950*, ed. Fay Bound Alberti. Palgrave Macmillan.

Bound Alberti, Fay. 2006b. Introduction: Emotion Theory and Medical History. In *Medicine, Emotion and Disease 1700-1950*, ed. Fay Bound Alberti. Palgrave Macmillan.

Bourdieu, Pierre. 1977. *Outline of a Theory of Practice*. Cambridge University Press.

Bowers, Rick. 2003. Roger Crab: Opposition Hunger Artist in 1650s England. *The Seventeenth Century* 18 (1): 93–112. https://doi.org/10.1080/0268117X.2003.10555521.

Breitenberg, Mark. 1996. *Anxious Masculinity in Early Modern England*. Cambridge University Press.

Bright, Timothie. 1586. *A Treatise of Melancholie, Containing the Causes Thereof, & Reasons of the Strange Effects It Worketh in Our Minds and Bodies: With the Physicke Cure, and Spirituall Consolation for Such As Haue Thereto Adioyned an Afflicted Conscience*. Thomas Vautrollier.

Brooke, Christopher. 2012. *Philosophic Pride: Stoicism and Political Thought from Lipsius to Rousseau*. Princeton University Press.

Brossard, Baptiste. 2018. *Why Do We Hurt Ourselves?: Understanding Self-Harm in Social Life*. Indiana University Press.

Browne, John. 1678. *A Compleat Discourse of Wounds, Both in General and Particular*. E. Flesher for William Jacob.

Brumberg, Joan Jacobs. 1988. *Fasting Girls: The Emergency of Anorexia Nervosa as a Modern Disease*. Harvard University Press.

Bugbee, John. 2019. Chaucer's Lucretia and What Augustine Really Said About Rape: Two Reconsiderations. *Traditio* 74:335–373.

Bulwer, John. 1644. *Chironomia, or, the Art of Manuall Rhetorique*. Tho(mas) Harper.

Bulwer, John. 1653. *Anthropometamorphosis: = Man Transform'd: Or, the Artificiall Changling*. William Hunt.

Burton, Robert. 1621. *The Anatomy of Melancholy*. John Lichfield and James Short, for Henry Cripps.

Bynum, Caroline Walker. 1987. *Holy Feast and Holy Fast: The Religious Significance of Food to Medieval Women*. University of California Press.

Cadman, Daniel John. 2011. Republicanism and Stoicism in Renaissance and Neo-Senecan Drama. PhD diss., Sheffield Hallam University.

Campana, Joseph. 2012. *The Pain of Reformation: Spenser, Vulnerability, and the Ethics of Masculinity*. Fordham University Press.

Carpi, Daniela. 2015. Caesar's Body in Shakespeare's Julius Caesar: Sacralization and De-Sacralization of Power. *Pólemos* 9 (2): 281–294. https://doi.org/10.1515/pol-2015-0019.

Carrera, Elena. 2019. Augustinian, Aristotelian, and Humanist Shaping of Medieval and Early Modern Emotion: Affectus, Afectio, and 'Affection' as Travelling Concepts. In *Before Emotion: The Language of Feeling, 400-1800*, ed. Michael W. Champion, Kirk Essary, and Juanita Feros Ruys. Routledge.

Caryl, Joseph. 1647. *An Exposition with Practicall Observations Continued upon the Eigth, Ninth and Tenth Chapters of the Book of Job*. A. Miller for Henry Overton ... and Luke Fawne ... and John Rothwell ... and Giles Calvert.

Celovsky, Lisa. 2005. Early Modern Masculinities and The Faerie Queene. *English Literary Renaissance* 35 (2): 210–247.

Chandler, Amy. 2012. Self-Injury as Embodied Emotion Work: Managing Rationality, Emotions and Bodies. *Sociology* 46 (3): 442–457. https://doi.org/10.1177/0038038511422589.

Chaney, Sarah. 2011a. 'A Hideous Torture on Himself': Madness and Self-Mutilation in Victorian Literature. *Journal of Medical Humanities* 32 (4): 279–289. https://doi.org/10.1007/s10912-011-9152-6.

Chaney, Sarah. 2011b. Self-Control, Selfishness and Mutilation: How 'Medical' is Self-Injury Anyway? *Medical History* 55 (3): 375–382.

Chaney, Sarah. 2017. *Psyche on the Skin: A History of Self-harm*. Reaktion Books.

Charland, Louis C., and R. S. White. 2015. Anatomy of a Passion: Shakespeare's The Winter's Tale as Case Study. In *Ordering Emotions in Europe, 1100-1800*, ed. Susan Broomhall. Brill.

Chaucer, Geoffrey. 2008. *The Riverside Chaucer*. Oxford University Press.

Cinpoeş, Nicoleta. 2018. Introduction. In *Doing Kyd: Essays on The Spanish Tragedy*, ed. Nicoleta Cinpoeş. Manchester University Press.

Clark, Andy. 2013. Gesture as Thought? In *The Hand, an Organ of the Mind: What the Manual Tells the Mental*, ed. Zdravko Radman. MIT Press.

Cock, Emily. 2013. Exchanging Flesh: Prostitution and Plastic Surgery in Seventeenth-Century England. PhD diss., University of Adelaide.

Cockcroft, Robert. 2003. *Rhetorical Affect in Early Modern Writing: Renaissance Passions Reconsidered*. Palgrave Macmillan UK.

Cohen, Esther. 2009. *The Modulated Scream: Pain in Late Medieval Culture*. University of Chicago Press.

Cohen, Michelle, and Tim Hitchcock. 2014. Introduction. In *English Masculinities, 1660-1800*, ed. Tim Hitchcock and Michelle Cohen. Routledge.

Coleman, Charley. 2022. OFCOM (Duty Regarding Prevention of Serious Self-Harm and Suicide) Bill [HL]: HL Bill 18 of 2022-23. https://lordslibrary.parliament.uk/research-briefings/lln-2022-0041/. Accessed 8 September 2022.

Cooper, Thomas. 1578. *Thesaurus Linguæ Romanæ & Britannicæ*. Henry Denham.

Cooper, Farah Karim. 2016. *The Hand on the Shakespearean Stage: Gesture, Touch and the Spectacle of Dismemberment*. Bloomsbury Publishing.

Corbett, Theodore G. 1975. The Cult of Lipsius: A Leading Source of Early Modern Spanish Statecraft. *Journal of the History of Ideas* 36 (1): 139–152. https://www.jstor.org/stable/2709016.

Craik, Katherine A., and Tanya Pollard. 2013. *Shakespearean Sensations: Experiencing Literature in Early Modern England*. Cambridge University Press.

Crooke, Helkiah. 1615. *Mikrokosmographia a Description of the Body of Man. Together With the Controversies Thereto Belonging*. William Jaggard.

Cuffari, Elena. 2012. Gestural Sense-Making: Hand Gestures as Intersubjective Linguistic Enactments. *Phenomenology and the Cognitive Sciences* 11 (4): 599–622. https://doi.org/10.1007/s11097-011-9244-9.

Daniel, Drew. 2013. *The Melancholy Assemblage: Affect and Epistemology in the English Renaissance*. Fordham University Press.

Daniel, Drew. 2022. *Joy of the Worm: Suicide and Pleasure in Early Modern English Literature*. Chicago University Press.

Davies, Jonathan. 2013. *Aspects of Violence in Renaissance Europe*. Ashgate Publishing Ltd.

Davis, Lloyd. 1993. *Guise and Disguise: Rhetoric and Characterization in the English Renaissance*. University of Toronto Press.

Dawson, Anthony. 2003. Introduction. In *Tamburlaine The Great: Part Two*, ed. Christopher Marlowe. Bloomsbury Publishing.

Dawson, Hannah. 2019. Shame in Early Modern Thought: From Sin to Sociability. *History of European Ideas* 45 (3): 377–398. https://doi.org/10.1080/01916599.2018.1534447.

De Marconville, Jean. 2012. A Treatise of the Good and Evell Tongue (Translated by T. S. (London: 1592)). In *The Unruly Tongue in Early Modern England: Three Treatises*, ed. Nathalie Vienne-Guerrin. Fairleigh Dickinson University Press.

Defoe, Daniel. 1724–1727. *A Tour Thro' the Whole Island of Great Britain, Divided into Circuits or Journies. Giving a Particular and Diverting Account of Whatever Is Curious and Worth Observation*. G. Strahan, W. Mears, R. Francklin, S. Chapman, R. Stagg, and J. Graves.

Di Cara, Nina H., Oliver S. P. Davis, Claire M. A. Haworth, Luke Sloan, and Lizzy Winstone. 2002. The Mental Health and Well-Being Profile of Young Adults Using Social Media. *Npj Mental Health Research* 1 (1): 1–11. https://doi.org/10.1038/s44184-022-00011-w.

Dijkhuizen, Jan Frans van. 2008. Partakers of Pain: Religious Meanings of Pain in Early Modern England. In *The Sense of Suffering: Constructions of Physical Pain in Early Modern Culture*, ed. Jan Frans van Dijkhuizen and Karl A. E. Enenkel. Brill.

Dijkhuizen, Van, Jan Frans, and Karl A. E. Enenkel. 2008. Introduction: Constructions of Physical Pain in Early Modern Culture. In *The Sense of Suffering: Constructions of Physical Pain in Early Modern Culture*, ed. Jan Frans van Dijkhuizen and Karl A. E. Enenkel. Brill.

Dixon, Robert. 1683. *Canidia, or, The Witches a Rhapsody*. S. Roycroft.

Dixon, Thomas. 2003. *From Passions to Emotions: The Creation of a Secular Psychological Category*. Cambridge University Press.

Dixon, Thomas. 2015. *Weeping Britannia: Portrait of a Nation in Tears*. Oxford University Press.

Donne, John. 1644. *Biathanatos a Declaration of That Paradoxe or Thesis, That Selfe-Homicide Is Not So Naturally Sinne, That It May Never Be Otherwise: Wherein the Nature and the Extent of All Those Lawes, Which Seeme to Be Violated by This Act, Are Diligently Surveyed*. Printed by John Dawson.

Drayton, Michael. 1622. *The Second Part, or a Continuance of Poly-Olbion from the Eighteenth Song Containing All the Tracts, Rivers, Mountaines, and Forrests: Intermixed with the Most Remarkable Stories, Antiquities, Wonders, Rarities, Pleasures, and Commodities of the East, and Northerne Parts of This Isle, Lying Betwixt the Two Famous Riuers of Thames, and Tweed*. Printed by Augustine Mathewes for John Marriott, John Grismand, and Thomas Dewe.

Du Vair, Guillaume. 1622. *A Buckler Against Adversitie, or, A Treatise of Constancie*. Bernard Alsop.

Dué, Casey. 2006. *The Captive Woman's Lament in Greek Tragedy*. University of Texas Press.

Duxfield, Andrew. 2015. *Christopher Marlowe and the Failure to Unify*. Ashgate Publishing Ltd.

Edwards, Catherine. 2005. Modelling Roman Suicide? The Afterlife of Cato. *Economy and Society* 34 (2): 200–222. https://www.tandfonline.com/doi/full/10.1080/0308514050054578.

Enterline, Lynn. 2011. *Shakespeare's Schoolroom: Rhetoric, Discipline, Emotion*. University of Pennsylvania Press.

Euripides. 1891. Trojan Women. In *The Plays of Euripides*, trans. E. P. Coleridge. Vol. 1 of 2. George Bell and Sons.

Favazza, Armando R. 1987. *Bodies Under Siege: Self-mutilation and Body Modification in Culture and Psychiatry*. John Hopkins University Press.

Fissell, Mary Elizabeth. 2004. *Vernacular Bodies: The Politics of Reproduction in Early Modern England*. Oxford University Press.

Floyd-Wilson, Mary, Matthew Greenfield, Gail Kern Paster, Tanya Pollard, Katherine Rowe, and Julian Yates. 2005. Shakespeare and Embodiment: An E-Conversation. *Literature Compass* 2 (1): 1–13.

Floyd-Wilson, Mary, and Garrett A. Sullivan Jr. 2007. Introduction: Inhabiting the Body, Inhabiting the World. In *Environment and Embodiment in Early*

Modern England, ed. Mary Floyd-Wilson and Garrett A. Sullivan Jr. Palgrave Macmillan.

Francis of Sales. 1871. *Introduction to the Devout Life*. Rivington.

Fraunce, Abraham. 1588. *The Arcadian Rhetorike: or The praecepts of rhetorike made plaine by Examples*. Thomas Orwin.

Freccero, Carla, Madhavi Menon, and Valerie Traub. 2013. Historicism and Unhistoricism in Queer Studies. *PMLA* 128 (3): 781–786. https://www.jstor.org/stable/23489323.

Freud, Sigmund. 2015. *Beyond the Pleasure Principle*. Dover Publications, Incorporated.

Frey, Christopher, and Leanore Lieblein. 2004. 'My Breasts Sear'd': The Self-Starved Female Body and A Woman Killed with Kindness. *Early Theatre* 7 (1): 45–67.

Fuller, Thomas. 1662. *The History of the Worthies of England*. J.G.W.L. and W.G. for Thomas Williams.

Gee, John. 1624. *The Foot out of the Snare: With a Detection of Sundry Late Practices and Impostures of the Priests and Jesuits in England*. H. Lownes for Robert Milbourn.

Geoffrey, Miles. 1996. *Shakespeare and the Constant Romans*. Oxford University Press.

Ghose, Indira. 2008. *Shakespeare and Laughter: A Cultural History*. Manchester University Press.

Goldberg, Jonathan, and Madhavi Menon. 2005. Queering History. *PMLA* 120 (5): 1608–1617.

Goodland, Katharine. 1999. 'Why Should Calamity Be Full of Words?': The Representation of Women's Laments for the Dead in Medieval English and Shakespearean Drama. PhD diss., Purdue University. ProQuest.

Gosson, Stephen. 1582. *Playes Confuted in Five Actions*. Thomas Gosson.

Goth, Maik. 2012. 'Killing, Hewing, Stabbing, Dagger-Drawing, Fighting, Butchery': Skin Penetration in Renaissance Tragedy and Its Bearing on Dramatic Theory. *Comparative Drama* 46 (2): 139–162.

Gratz, Kim L., Alexander L. Chapman, Katherine L. Dixon-Gordon, and Matthew T. Tull. 2015. Diagnosis and Characterization of DSM-5 Nonsuicidal Self-Injury Disorder Using the Clinician-Administered Nonsuicidal Self-Injury Disorder Index. *Assessment* 22 (5): 527–539. https://doi.org/10.1177/1073191114565878.

Greenblatt, Stephen. 1997. Mutilation and Meaning. In *The Body in Parts: Fantasies of Corporeality in Early Modern Europe*, ed. David A. Hillman and Carla Mazzio. Routledge.

Greene, Robert. 1583. *Perimedes the Blacke-Smith a Golden Methode, How to use the Minde in Pleasant and Profitable Exercise*. John Wolfe, for Edward White.

Greene, John. 1615. *A Refutation of the Apology for Actors*. W. White.

Greenfield, Matthew. 2004. Christopher Marlowe's Wound Knowledge. *PMLA* 119 (2): 233–246. https://doi.org/10.1632/003081204X21289.

Gregory, Edmund. 1646. *An Historical Anatomy of Christian Melancholy, Sympathetically Set Forth, in a Threefold State of the Soul. 1 Endued with Grace, 2 Ensnared in Sin, 3 Troubled in Conscience*. Printed for Humphrey Moseley.

Groebner, Valentin. 1995. Losing Face, Saving Face: Noses and Honour in the Late Medieval Town. *History Workshop Journal* 40 (1): 1–15. https://doi.org/10.1093/hwj/40.1.1.

Groschwitz, Rebecca C., Michael Kaess, Gloria Fischer, et al. 2015. The Association of Non-Suicidal Self-Injury and Suicidal Behavior According to DSM-5 in Adolescent Psychiatric Inpatients. *Psychiatry Research* 228 (3): 454–461. https://doi.org/10.1016/j.psychres.2015.06.019.

Gutierrez, Nancy A. 2016. *'Shall She Famish Then?': Female Food Refusal in Early Modern England*. Routledge.

H.A. 1669. *Mirabile Pecci, or, The Non-Such Wonder of the Peak in Darby-shire*. T. Parkhurst … and G. Calvert.

Hall, Jon. 2004. Cicero and Quintilian on the Oratorical Use of Hand Gestures. *The Classical Quarterly* 54 (1): 143–160. https://www.jstor.org/stable/3556289.

Hamilton, A. C., Shohachi Fukuda, Toshiyuki Suzuki, and Hiroshi Yamashita, eds. 2014. *Spenser: The Faerie Queene*. 2nd ed. Routledge. https://doi.org/10.4324/9781315834696.

Hanson, Elizabeth. 1991. Torture and Truth in Renaissance England. *Representations* 34:53–84. https://www.jstor.org/stable/2928770.

Hardy, Nathaniel. 1666. *Lamentation, Mourning, and Woe Sighed Forth in a Sermon Preached in the Parish-Church of St. Martin in the Fields, on the 9th Day of September: Being the next Lords-Day after the Dismal Fire in the City of London*. Thomas Newcomb.

Hattaway, Michael. 2005. The Spanish Tragedy: Architectonic Design. In *Elizabethan Popular Theatre: Plays in Performance*. Routledge.

Heath, James. 1689. *Englands Chronicle, or, The Lives & Reigns of the Kings and Queens from the Time of Julius Cæsar to the Present Reign of K. William and Q. Mary Containing the Remarkable Transactions and Revolutions in Peace and War, Both at Home and Abroad, as they Relate to this Kingdom, with the Wars, Policies, Religion and Customs, Success and Misfortunes as Well of the Ancient Britains, as Roman, Saxon, Danish, and Norman Conquerors, with Copper Cuts and Whatever Else Is Conduceable to the Illustration of History*. Printed for Benj. Crayle, N. Bodington and G. Conyer.

Heilbron, Nicole, Jill Compton, Stephanie Daniel, and David Goldston. 2010. The Problematic Label of Suicide Gesture. *Professional Psychology: Research and Practice* 41 (3): 221–227.

Heyerick, Isabelle. 2019. Is There an I in Impact? Considering the Two-Way Process of Public Engagement. *Exchanges: The Interdisciplinary Research Journal* 7 (1): 82–93. https://doi.org/10.31273/eirj.v7i1.520.

Heywood, Thomas. 1612. *An Apology for Actors.* Nicholas Okes.

Heywood, Thomas. 1624. *Gynaikeion: or, Nine Books of Various History. Concerninge Women Inscribed by The Names of Ye Nine Muses.* Adam Islip.

Hinz, Andreas, and Aglaja Stirn. 2008. Tattoos, Body Piercings, and Self-Injury: Is There a Connection? Investigations on a Core Group of Participants Practicing Body Modification. *Psychotherapy Research* 18 (3): 326–333. https://doi.org/10.1080/10503300701506938.

Hobbes, Thomas. 1994. Letter 183 20 [30 October 1668] Thomas Hobbes to John Brooke, from Chatsworth. In *The Clarendon Edition of the Works of Thomas Hobbes, Vol. 7: The Correspondence, Vol 2. 1660–1679*, ed. Noel Malcolm. Oxford University Press.

Homer. 1924. *The Iliad.* Translated by A. T. Murray. Heinemann.

Hopkins, Lisa, and Sean McEvoy. 2008. *Christopher Marlowe, Renaissance Dramatist.* Edinburgh University Press.

Huckins-Noss, Andrew M., Kathryn R. Fox, Alexander J. Millner, Matthew K. Nock, and Jessica Stubbing. 2023. The Classification of Suicide Gestures: Examining the Validity, Measurement, and Level of Intent of a Controversial Construct. *Suicide and Life-Threatening Behaviour* 53 (2): 188–197.

Huth, Kimberly. 2009. Figures of Pain: Suffering and Selfhood in Early Modern English Literature. PhD diss., The University of Wisconsin – Madison. ProQuest.

Hutson, Lorna. 2005. Rethinking the 'Spectacle of the Scaffold': Juridical Epistemologies and English Revenge Tragedy. *Representations* 89 (1): 30–58. https://www.jstor.org/stable/10.1525/rep.2005.89.1.30.

Iannaccone, Mara, Stefania Cella, Paolo Cotrufo, Federico Manzi, Sabato Antonio Manzi, and Luigia Visconti. 2013. My Body and Me: Self-Injurious Behaviors and Body Modifications in Eating Disorders-Preliminary Results. *Eating Disorders* 21 (2): 130–139. https://doi.org/10.1080/10640266.2013.761087.

Ibañez, Lindsey M., and Jim D. Taylor. 2015. Sociological Approaches to Self-Injury. *Sociology Compass* 9 (12): 1005–1014.

Imogen Knox. 2023. Pin-Swallowing and Self-Destruction in Early Modern British and Irish Supernatural Narratives. *Cultural and Social History*, 20 (4): 499–515.

Inckle, Kay. 2010. Telling Tales? Using Ethnographic Fictions to Speak Embodied 'Truth'. *Qualitative Research* 10 (1): 27–47. https://doi.org/10.1177/1468794109348681.

Ingram, Martin. 1984. Ridings, Rough Music and the 'Reform of Popular Culture' in Early Modern England. *Past & Present* 105:79–113. https://www.jstor.org/stable/650546.

Irish, David. 1700. *Levamen Infirmi: Or, Cordial Counsel to the Sick and Diseased.*

Irvine, Martin. 1997. Abelard and (Re)Writing the Male Body: Castration, Identity, and Remasculinization. In *Becoming Male in the Middle Ages*, ed. Jeffrey Jerome Cohen and Bonnie Wheeler. Garland Publishing.

James, Susan. 1999. *Passion and Action: The Emotions in Seventeenth-Century Philosophy.* Oxford University Press.

Jonson, Ben. 1756. Timber, or Discoveries Made Upon Men and Matter. In *The Works of Ben. Johnson.* Printed for D. Midwinter; W. Innys and J. Richardson; J. Knapton; T. Wotton; C. Hitch and L. Hawes.

Kahn, Coppélia. 1997. *Roman Shakespeare: Warriors, Wounds and Women.* Taylor & Francis Group.

Kaku, Archana. 2020. Burning the Body: The Bodily Politics of Tibetan Self-Immolation. *Theory & Event* 23 (3): 573–606.

Kendon, Adam. 2004. *Gesture: Visible Action as Utterance.* Cambridge University Press.

King, John N. 2006. Introduction. In *Foxe's 'Book of Martyrs' and Early Modern Print Culture*, ed. John N. King. Cambridge University Press.

Knox, Kerry L., and Janis Whitlock. 2007. The Relationship Between Self-Injurious Behavior and Suicide in a Young Adult Population. *Archives of Pediatrics & Adolescent Medicine* 161 (7): 634–640. https://doi.org/10.1001/archpedi.161.7.634.

Kristeva, Julia. 1982. *Powers of Horror: An Essay on Abjection.* Translated by Leon S. Roudiez. Columbia University Press.

Kuchta, David. 2002. *The Three-Piece Suit and Modern Masculinity: England, 1550-1850.* University of California Press.

L.S. 1652. *Natures Dowrie: Or The Peoples Native Liberty Asserted.* W.R. at the signe of the Vnicorn in Pauls Church-yard.

La Calprenède, Gaultier de Coste. 1674. *Hymen's Præludia, or Loves Master-Piece Being That so Much Admired Romance, Intituled Cleopatra: In Twelve Parts.* W. R. and J. R.

La Charriere, Joseph de. 1712 *A Treatise of the Operations of Surgery. Wherein Are Mechanically Explain'd the Causes of the Diseases in Which They Are Needful, … To Which Is Added, a Treatise of Wounds, … Translated from the Third Edition of the French, Enlarg'd, Corrected and Revis'd by the Author, Joseph de la Charriere.* Printed by R. Brugis, for D. Brown, and W. Mears; and T. Ballard.

Langley, Eric Francis. 2009. *Narcissism and Suicide in Shakespeare and His Contemporaries.* Oxford University Press.

LeClerc, M. 1696. *The Compleat Surgeon, or, the Whole Art of Surgery Explain'd in a Most Familiar Method. Containing an Exact Account of Its Principles and Several Parts, viz. Of the Bones, Muscles, Tumurs, Ulcers, and Wounds Simple and Complicated, or Those by Gun-shot; as Also of Venereal Diseases, the Scruvy, Fractures, Luxations, and All Sorts of Chirurgical Operations; Together With*

Their Proper Bandages and Dressings. To Which Is Added, a Chirurgical Dispensatory; Shewing the Manner How to Prepare All Such Medicines as Are Necessary for a Surgeon, and Particularly the Mercurial Panacæa. Printed for M. Gillyflower, in Westminster-Hall; T. Goodwin, and M. Wotton, in Fleet-street; J. Walthoe, in the Middle-Temple Cloyster; and R. Parker, under the Royal-Exchange, in Cornhill.

Lipsius, Justus. 1595. *Two Bookes of Constancie ... Containing, Principallie, A Comfortable Conference, in Common Calamities. And Will Serue for a Singular Consolation to All That Are Priuately Distressed, of Afflicted, Either in Body or Mind*. Translated by John Stradling. P. Short for Richard Johnes.

Lodhia, Sheetal. 2009. 'The House Is Hers, the Soul Is But a Tenant': Material Self-Fashioning and Revenge Tragedy. *Early Theatre* 12 (2): 135–161.

Long, A. A. 2003. Stoicism in the Philosophical Tradition: Spinoza, Lipsius, Butler. In *The Cambridge Companion to the Stoics*, ed. Brad Inwood. Cambridge University Press.

Lye, Thomas. 1660. *The King of Terrors Metamorphosis. Or, Death Transform'd into Sleep. A Sermon Preached at the Funeral of Mrs. Elizabeth Nicoll, Daughter to That Worthy, Eminently Pious, and Charitable Citizen of London, Mr. John Walter Deceased, and Late Wife of Mr. William Nicoll of London Draper*. Printed by M.S. for Hen. Cripps.

MacDonald, Michael, and Terence R. Murphy. 1993. *Sleepless Souls: Suicide in Early Modern England*. Clarendon Press.

Magnusson, Lynne. 2016. Shakespearean Tragedy and the Language of Lament. In *The Oxford Handbook of Shakespearean Tragedy*. Oxford University Press.

Mandeville, Sir John. 1582. *The Voyages and Trauailes of Sir John Maundeuile Knight Wherein Is Treated of the Way Towards Hierusalem, and of the Meruailes of Inde, with Other Lands and Countries*. Thomas Este.

Marshall, Cynthia. 1994. Portia's Wound, Calphurnia's Dream: Reading Character in Julius Caesar. *English Literary Renaissance* 24 (2): 471–488.

Marshall, Claire. 2000. The Politics of Self-Mutilation: Forms of Female Devotion in the Late Middle Ages. In *The Body in Late Medieval and Early Modern Culture*, ed. Darryll Grantley and Nina Taunton. Ashgate Publishing.

Marshall, Cynthia. 2022. *The Shattering of the Self: Violence, Subjectivity, and Early Modern Texts*. John Hopkins University Press.

Martin, Matthew R. 2016. *Tragedy and Trauma in the Plays of Christopher Marlowe*. Routledge.

Martindale, Charles, and Michelle Martindale. 1994. *Shakespeare and the Uses of Antiquity An Introductory Essay*. Taylor & Francis Group.

Maus, Katharine Eisaman. 1986. Taking Tropes Seriously: Language and Violence in Shakespeare's Rape of Lucrece. *Shakespeare Quarterly* 37 (1): 66–82. https://doi.org/10.2307/2870192.

Mayhew, Jenny. 2008. Godly Beds of Pain: Pain in English Protestant Manuals (ca. 1550-1650). In *The Sense of Suffering: Constructions of Physical Pain in Early Modern Culture*, ed. Jan Frans van Dijkhuizen and Karl A. E. Enenkel. Brill.

Mayne, Jasper. 1648. *The Amorous Warre*. S.N.

Mazzio, Carla. 1998. Sins of the Tongue in Early Modern England. *Modern Language Studies* 28 (3/4): 95–124. https://www.jstor.org/stable/3195467.

Mbembé, Achille. 2003. "Necropolitics", Translated by Libby Meintjes. *Public Culture* 15 (1): 11–40.

McDonald, Russ. 2017. Rhetoric and Theater. In *The Oxford Handbook of Rhetorical Studies*. Oxford University Press.

McGuckin, John Anthony. 2004. *The Westminster Handbook to Origen*. Westminster John Knox Press.

McIntosh, Hugh. 2010. The Social Masochism of Shakespeare's Sonnets. *Studies in English Literature, 1500-1900* 50 (1): 109–125. https://www.jstor.org/stable/40658422.

Menon, Madhavi. 2013. Historicism and Unhistoricism in Queer Studies. *PMLA* 128 (3): 782–784. https://doi.org/10.1632/S003081290012317X.

Miles, Thomas. 1613. *The Treasurie of Auncient and Moderne Times*. W. Jaggard.

Millard, Chris. 2015. *A History of Self-Harm in Britain: A Genealogy of Cutting and Overdosing*. Palgrave Macmillan.

Miller, David Lee. 2015. The Voice of Caesar's Wounds: The Politics of Martyrdom in Shakespeare's Julius Caesar. In *Forms of Association: Making Publics in Early Modern Europe*, ed. Marlene Eberhart and Paul Yachnin. University of Massachusetts Press.

Millett, Bella, and Jocelyn Wogan-Browne. 1992. *Medieval English Prose for Women: Selections from the Katherine Group and Ancrene Wisse*. Oxford University Press.

Minois, Georges. 1999. *History of Suicide: Voluntary Death in Western Culture*. John Hopkins University Press.

Monta, Susannah Brietz. 2005. *Martyrdom and Literature in Early Modern England*. Cambridge University Press.

Montaigne, Michel de. 1595. Of the inconstancie of our actions. In *Essayes of Morall, Politike, and Millitarie Discourses*, trans. John Florio. Edward Blount.

Montaigne, Michel de. 1613. *Essays Written in French by Michael Lord of Montaigne*. Translated by John Florio. Melch. Bradwood for Edward Blount and William Barret.

More, Thomas. 1553. *A Dialoge of Comfort Against Tribulacion, Made by Syr Thomas More Knyght, and Set Foorth by the Name of an Hu[n]Garie[n], Not before This Time Imprinted*.

Moréri, Louis. 1694. *The Great Historical, Geographical and Poetical Dictionary Being a Curious Miscellany of Sacred and Prophane History.* Printed for Henry Rhodes, Luke Meredick and John Harris.

Moro-Abadia, Oscar. 2009. Thinking about 'Presentism' from a Historian's Perspective: Herbert Butterfield and Hélène Metzger. *History of Science* 47 (1): 55–77. https://doi.org/10.1177/007327530904700104.

Morris, Jeffrey B. 1997. To (Re)Fashion a Gentleman: Ralegh's Disgrace in Spenser's Legend of Courtesy. *Studies in Philology* 94 (1): 38–58. https://www.jstor.org/stable/4174567.

Moscoso, Javier. 2020. Learning and Teaching Pain. In *The Routledge History of Emotions in Europe, 1100-1700*, ed. Susan Broomhall and Andrew Lynch. Routledge.

Munday, Anthony. 1580. *A Second and Third Blast of Retrait from Plaies and Theaters.* Henrie Denham.

Murray, Jacqueline. 2006. Sexual Mutilation and Castration Anxiety: A Medieval Perspective. In *The Boswell Thesis: Essays on Christianity, Social Tolerance, and Homosexuality*, ed. Mathew Kuefler. University of Chicago Press.

Musica, Eughenii. 2018. Vindicating Revenge. In *Doing Kyd: Essays on The Spanish Tragedy*, ed. Nicoleta Cinpoeş. Manchester University Press.

N.H. 1694. *The Ladies Dictionary, Being a General Entertainment of the Fair-Sex.* John Dunton.

Nance, John V. 2013. Gross Anatomies: Mapping Matter and Literary Form in Thomas Nashe and Andreas Vesalius. In *The Age of Thomas Nashe*, ed. Stephen Guy-Bray, Joan Pon Linton, and Steve Mentz. Routledge.

Nazarian, Cynthia. 2016. Sympathy Wounds, Rivers of Blood: The Politics of Fellow Feeling in Spenser's Faerie Queene and a View of the State of Ireland. *Modern Philology* 113 (3): 331–352.

Neely, Carol Thomas. 2004. *Distracted Subjects: Madness and Gender in Shakespeare and Early Modern Culture.* Cornell University Press.

Newton, James. 1675. *They That Have Any Friends Distracted or Melancholy Are Desired to Accept This.* S.n.

Nicholson, Catherine. 2013. *Uncommon Tongues: Eloquence and Eccentricity in the English Renaissance.* University of Pennsylvania Press.

Old Bailey Proceedings Online. 2025a. (www.oldbaileyonline.org, version 9.0) *Ordinary of Newgate's Account* March 1741 (OA17410318). https://www.oldbaileyonline.org/record/OA17410318. Accessed 13 June 2025.

Old Bailey Proceedings Online. 2025b. (www.oldbaileyonline.org, version 9.0) *Ordinary of Newgate's Account* September 1731 (OA17310924). https://www.oldbaileyonline.org/record/OA17310924. Accessed 13 June 2025.

Old Bailey Proceedings Online. 2025c. (www.oldbaileyonline.org, version 9.0) *Ordinary of Newgate's Account* April 1721 (OA17210403). https://www.oldbaileyonline.org/record/OA17210403. Accessed 13 June 2025.

Oliver, Richard. 1700. *A Sermon Preach'd at the Funeral of George Payne*. Printed for Arthur Bettesworth and William Webb.

Owen, John. 1668. *Of the Mortification of Sin in Believers*. Nathanael Ponder, at the Peacock in the Poultrey near Cornhill, and in Chancery-Lane near Fleet-Street.

Owens, Margaret E. 2005. *Stages of Dismemberment: The Fragmented Body in Late Medieval and Early Modern Drama*. University of Delaware Press.

Oxford English Dictionary.https://www.oed.com/.

Pain, Rachel. 2014. Impact: Striking a Blow or Walking Together? *ACME: An International Journal for Critical Geographies* 13 (1): 19–23. https://doi.org/10.14288/acme.v13i1.986.

Panek, Jennifer. 2018. The Nice Valour's Anatomy of Shame. *English Literary Renaissance* 48 (3): 339–367. https://doi.org/10.1086/699030.

Paoli, Gio Antonio de. 1636. *The Lives of All the Roman Emperors*. N. and I. Okes.

Persons/Parsons, Robert. 1603. *An Examen of the Calendar or Catalogue of Protestant Saincts, Martyrs, and Consellors, Deuised by [J]on Fox, and Prefixed Before His Great Volume, of Acts and Monuments*.

Paster, Gail Kern. 1989. 'In the Spirit of Men There Is No Blood': Blood as Trope of Gender in Julius Caesar. *Shakespeare Quarterly* 40 (3): 284–298. https://doi.org/10.2307/2870724.

Paster, Gail Kern. 1993. *The Body Embarrassed: Drama and the Disciplines of Shame in Early Modern England*. Cornell University Press.

Patrick, Simon. 1671. *A Consolatory Discourse to Prevent Immoderate Grief for the Death of Our Friends*. R. W. for Francis Tyton.

Peacham, Henry. 1593. *The Garden of Eloquence, Conteining the Most Excellent Ornaments, Exornations, Lights, Flowers, and Formes of Speech, Commonly Called the Figures of Rhetoricke*. R.F. for H. Jackson.

Pearson, Meg. 2012. 'Raving, Impatient, Desperate, and Mad': Tamburlaine's Spectacular Collapse. *Marlowe Studies* 2:87–102.

Pender, Stephen. 2008. Seeing, Feeling, Judging: Pain in the Early Modern Imagination. In *The Sense of Suffering: Constructions of Physical Pain in Early Modern Culture*, ed. Jan Frans van Dijkhuizen and Karl A. E. Enenkel. Brill.

Pender, Stephen. 2010. Rhetoric, Grief, and the Imagination in Early Modern England. *Philosophy & Rhetoric* 43 (1): 54–85. https://doi.org/10.5325/philrhet.43.1.0054.

Perkell, Christine. 2008. Reading the Laments of Iliad 24. In *Lament: Studies in the Ancient Mediterranean and Beyond*, ed. Ann Suter. Oxford University Press, Incorporated.

Perkins, William. 1625. *A Salve for a Sicke Man: Or, A Treatise Containing the Nature, Differences, and Kinds of Death; as Also the Right Manner of Dying Well*. John Legatt, dwelling in Little-Wood-Streete (first published 1595).

Perkins, William. 2012. A Direction for the Government of the Tongue According to Gods Worde (Cambridge: 1595). In *The Unruly Tongue in Early Modern England: Three Treatises*, ed. Nathalie Vienne-Guerrin. Fairleigh Dickinson University Press.

Pertile, Giulio J. 2016. King Lear and the Uses of Mortification. *Shakespeare Quarterly* 67 (3): 319–343. https://doi.org/10.1353/shq.2016.0039.

Philipot, Thomas. 1674. *Self-Homicide-Murther, or, Some Antidotes and Arguments Gleaned out of the Treasuries of Our Modern Casuists and Divines against That Horrid and Reigning Sin of Self-Murther*. W. Downing for Edward Thomas.

Pierce, Ezra. 1692. *A Discourse of Self-Murder*. Rich. Baldwin.

Pilkington, Laetitia. 1748. *Memoirs of Mrs. Lætitia Pilkington, Wife to the Rev. Mr. Matthew Pilkington*.

Plutarch. 1603. *The Philosophie, Commonlie Called, the Morals Written by the Learned Philosopher Plutarch of Chæronea. Translated Out of Greeke into English, and Conferred with the Latine Translations and the French, by Philemon Holland of Coventrie, Doctor in Physicke*. Translated by Philemon Holland. Arnold Hatfield.

Plutarch. 2011. *Plutarch's Lives, Volume 2*. Random House Publishing Group.

Pollard, Tanya. 2010. Tragedy and Revenge. In *The Oxford Companion to English Renaissance Tragedy*, ed. Emma Smith and Garrett A. Sullivan Jr. Cambridge University Press.

Pollard, Tanya. 2012. What's Hecuba to Shakespeare? *Renaissance Quarterly* 65 (4): 1060–1093. https://doi.org/10.1086/669345.

Pope, Alexander. 1735. *Mr. Pope's Literary Correspondence, Volume the Third. With Letters To, and From, the Duke of Shrewberry, Dean Swift, Lord Lansdowne, Lady Chudleigh, Bishop of St. Asaph, Mrs. Manley, Sir Berkeley Lucy, Mrs Thomas, &c*. E. Curll.

Porter, Jerome. 1632. *The Flowers of the Liues of the Most Renowned Saincts of the Three Kingdoms England Scotland, and Ireland Written and Collected out of the Best Authours and Manuscripts of Our Nation, and Distributed According to Their Feasts in the Calendar*.

Powell, Hilary. 2022. Voices in Psychosis: A Medieval Perspective. In *Voices in Psychosis: Interdisciplinary Perspectives*, ed. Ben Alderson-Day, Charles Fernyhough, and Angela Woods. Oxford University Press.

Prynne, William. 1659. *Historiarchos, or, The Exact Recorder Being the Most Faithfull Remembrancer of the Most Remarkable Transactions of Estate and of All the English Lawes*. Printed for Francis Coles.

Puttenham, George. 1589. *The Arte of English Poesie*. Richard Field.

R. W. 1695. *An Essay on Grief: With the Causes and Remedies of It*. Printed by Leon Lichfield.

Raleigh, Walter. 1848. *Discovery of the Large, Rich and Beautiful Empire of Guiana, by Sir W. Ralegh: With a Relation of Great and Golden City of Manoa (Which the Spaniards call El Dorado), etc. (1596)*. Ashgate Publishing Ltd.

Reynolds, John. 1669. *A Discourse Upon Prodigious Abstinence: Occasioned by the Twelve Months Fasting of Martha Taylor, The Famed Derbyshire Damosell: Proving That Without Any Miracle, the Texture of Humane Bodes My be So altered, that Lafe May Be Long Continued Without the Supplies of Meat & Drink.* R.W. for Nevill Simmons … and Dorman Newman.

Riehl, Anna. 2009. Eying the Thought Awry: The Anamorphosis of John Donne's Poetry. *English Literary Renaissance* 39 (1): 141–162.

Roach, Joseph R. 1985. *The Player's Passion: Studies in the Science of Acting*. University of Delaware Press.

Robin the Plow-Man's Courage; OR, Harry the Millers Miserable Misfortune in Courting of Young Kate, Who Declared He Had Lost His Testicles, and Therefore Far Unfit for Wedlock. 1675–1696. Document. London: Printed for P. Brooksby, J. Deacon, J. Blare, and J. Back. From UCSB English Broadside Ballad Archive. https://ebba.english.ucsb.edu/ballad/21321.

Robins, Thomas. 1669. *The Wonder of the World*. Tho[mas] Passinger.

Robinson, Benedict S. 2017. Thinking Feeling. In *Affect Theory and Early Modern Texts: Politics, Ecologies, and Form*, ed. Amanda Bailey and Mario DiGangi. Springer.

Rogers, Thomas. 1576. *A Philosophicall Discourse, Entituled, The Anatomie of the Minde*. John Charlewood.

Rossini, N. 2012. *Reinterpreting Gesture as Language: Language "in Action"*. IOS Press.

Roulston, Chris. 2013. New Approaches to the Queer 18th Century. *Literature Compass* 10 (10): 761–770. https://doi.org/10.1111/lic3.12103.

Rowe, Katherine. 1997. God's Handy Worke. In *The Body in Parts: Fantasies of Corporeality in Early Modern Europe*, ed. David A. Hillman and Carla Mazzio. Routledge.

Ryder, Hugh. 1685. *New Practical Observations in Surgery Containing Divers Remarkable Cases and Cures*. James Partridge.

Samuelsen, Helle, and Vibeke Steffen. 2004. The Relevance of Foucault and Bourdieu for Medical Anthropology. *Anthropology and Medicine* 11 (1): 3–10.

Sandlin, Jennifer A., Jake Burdick, and Emma Rich. 2017. Problematizing Public Engagement within Public Pedagogy Research and Practice. *Discourse: Studies in the Cultural Politics of Education* 38 (6): 823–835. https://doi.org/10.1080/01596306.2016.1196343.

Saunders, Rebecca. 2007. *Lamentation and Modernity in Literature, Philosophy, and Culture*. Springer.

Sawday, Jonathan. 2013. *The Body Emblazoned: Dissection and the Human Body in Renaissance Culture*. Routledge.

Scarry, Elaine. 1988. *The Body in Pain: The Making and Unmaking of the World.* Oxford University Press.

Scheer, Monique. 2012. Are Emotions a Kind of Practice (and Is That What Makes Them Have a History)? A Bourdieuian Approach to Understanding Emotion. *History and Theory 51* (2): 193–220.

Schmidt, Jeremy. 2016. *Melancholy and the Care of the Soul: Religion, Moral Philosophy and Madness in Early Modern England.* Routledge.

Schneider, Robert A. 1986. Mortification on Parade: Penitential Processions in Sixteenth- and Seventeenth-Century France. *Renaissance and Reformation* 22 (1): 123–146.

Schoenfeldt, Michael C. 1999. *Bodies and Selves in Early Modern England: Physiology and Inwardness in Spenser, Shakespeare, Herbert and Milton.* Cambridge University Press.

Schwarz, Kathryn. 2000. *Tough Love: Amazon Encounters in the English Renaissance.* Duke University Press.

Searle, Alison. 2018. 'A Kind of Agonie in My Thoughts': Writing Puritan and Non-Conformist Women's Pain in 17th-Century England. *Medical Humanities* 44 (2): 125–136. https://doi.org/10.1136/medhum-2017-011407.

Sellberg, Karin, and Lena Wånggren. 2013. The Dismemberment of Will: Early Modern Fear of Castration. In *Castration and Culture in the Middle Ages*, ed. Larissa Tracy. Boydell and Brewer.

Seth, Sanjay. 2021. *Beyond Reason: Postcolonial Theory and the Social Sciences.* Oxford University Press.

Seton, Alexander. 1699. *A Treatise of Mutilation and Demembration.* Andrew Symson.

Shafte, J. 1673. *The Great Law of Nature, or, Self-Preservation Examined, Asserted and Vindicated from Mr. Hobbes His Abuses.* Printed by the author.

Shakespeare, William. 2005. *The Oxford Shakespeare: Complete Works.* 2nd ed. Edited by John Jowett, William Montgomery, Gary Taylor, and Stanley Wells. Oxford University Press.

Shakespeare, William. 2006. The Rape of Lucrece. In *The Poems: Venus and Adonis, The Rape of Lucrece, The Phoenix and the Turtle, The Passionate Pilgrim, A Lover's Complaint*, The New Cambridge Shakespeare, ed. John Roe, 2nd ed. Cambridge University Press. https://doi.org/10.1017/9780511819094.009.

Shakespeare, William. 2017. *Troilus and Cressida.* In *The New Cambridge Shakespeare*, ed. Anthony B. Dawson, 2nd ed. Cambridge University Press.

Sharp, Jane. 1671. *The Midwives Book.* Simon Miller.

Shepard, Alan. 2002. *Marlowe's Soldiers: Rhetorics of Masculinity in the Age of the Armada.* Ashgate Publishing.

Shepard, Alexandra. 2006. *Meanings of Manhood in Early Modern England.* Oxford University Press.

Shortslef, Emily. 2015. Weeping, Wailing, Sighing, Railing: Shakespeare and the Drama of Complaint. PhD diss., Columbia University.

Silverman, Lisa. 2001. *Tortured Subjects: Pain, Truth, and the Body in Early Modern France*. University of Chicago Press.

Skinner, Patricia. 2014. The Gendered Nose and its Lack: 'Medieval' Nose-Cutting and its Modern Manifestations. *Journal of Women's History* 26 (1): 45–67.

Skuse, Alanna. 2017. Missing Parts in The Shoemaker's Holiday. *Renaissance Drama* 45 (2): 161–179. https://doi.org/10.1086/694329.

Skuse, Alanna. 2020. 'One Stroak of His Razour': Tales of Self Gelding in Early Modern England. *Social History of Medicine* 33 (2): 377–393. https://doi.org/10.1093/shm/hky100.

Skuse, Alanna. 2021. *Surgery and Selfhood in Early Modern England: Altered Bodies and Contexts of Identity*. Cambridge University Press.

Slatyer, William. 1621. *The History of Great Britanie from the First Peopling of This Island to This Present Raigne of or [Sic] Happy and Peacefull Monarke K: James*. W: Stansby, for Rich: Meighen.

Slatyer, William. 1643. *The Compleat Christian, and Compleat Armour and Armoury of a Christian, Fitting Him with All Necessary Furniture for That His Holy Profession, or, The Doctrine of Salvation*.

Smith, Richard. 1609. *The Prudentiall Balance of Religion Wherin the Catholike and Protestant Religion Are Weighed Together with the Weights of Prudence, and Right Reason*. François Bellet.

Smith, Lisa Wynne. 2008. 'An Account of an Unaccountable Distemper': The Experience of Pain in Early Eighteenth-Century England and France. *Eighteenth Century Studies* 41 (4): 459–480.

Solga, Kim. 2006. Rape's Metatheatrical Return: Rehearsing Sexual Violence among the Early Moderns. *Theatre Journal* 58 (1): 53–72. https://www.jstor.org/stable/25069779.

Somerset, Anne. 2021. *Unnatural Murder: Poison In The Court of James I*. Hachette UK.

Speller, John R. W. 2011. *Bourdieu and Literature*. Open Book Publishers.

Spelman, John. 1709. *The Life of Ælfred the Great, by Sir John Spelman Kt. From the Original Manuscript in the Bodlejan [sic] Library: With Considerable Additions*. Guil. Lancaster.

Spierenburg, Petrus Cornelis. 1984. *The Spectacle of Suffering: Executions and the Evolution of Repression: From a Preindustrial Metropolis to the European Experience*. Cambridge University Press.

Steenbergh, Kristine. 2008. Green Wounds: Pain, Anger and Revenge in Early Modern Culture. In *The Sense of Suffering: Constructions of Physical Pain in Early Modern Culture*, ed. Jan Frans van Dijkhuizen and Karl A. E. Enenkel. Brill.

Steggals, Peter. 2013. Making Sense of Self-Harm: Exploring the Cultural Meaning and Social Context of Non-Suicidal Self-Injury. PhD diss., Newcastle University. http://theses.ncl.ac.uk/jspui/handle/10443/2143.

Steggals, Peter, Steph Lawler, and Ruth Graham. 2020. The Social Life of Self-Injury: Exploring the Communicative Dimension of a Very Personal Practice. *Sociology of Health & Illness* 42 (1): 157–170. https://doi.org/10.1111/1467-9566.12994.

Stow, John. 1618. *The Abridgement of the English Chronicle, First Collected by M. John Stow, and After Him Augmented with Very Many Memorable Antiquities, and Continued with Matters Forreine and Domesticall, Unto the Beginning of the Yeare, 1618.* Edward Allde and Nicholas Okes for the Company of Stationers.

Strier, Richard. 2011. *The Unrepentant Renaissance: From Petrarch to Shakespeare to Milton.* University of Chicago Press.

Stubbes, Phillip. 1877. *The Anatomie of Abuses.* New Shakspere Society (first printed 1583).

Sullivan, Erin. 2013. A Disease unto Death: Sadness in the Time of Shakespeare. In *Emotions and Health 1200-1700*, ed. Elena Carrera. Brill.

Suter, Ann. 2008. Introduction. In *Lament: Studies in the Ancient Mediterranean and Beyond*, ed. Ann Suter. Oxford University Press, Incorporated.

Sym, John. 1637. *Lifes Preservative Against Self-Killing. Or, An Useful Treatise Concerning Life and Self-Murder.* M. Flesher, for R. Dawlman, and L. Fawne.

Taylor, Gary. 2000. *Castration: An Abbreviated History of Western Manhood.* Routledge Ltd.

THE / Quaker's Wives Lamentation / For the LOSS of / Her Husbands Jewels, / Who Gelded Himself (in Petticoat-Lane) to Vex His Wife. 1684–1700. Document. London: Printed for James Bissel, at the Bible and Harp in West-Smithfield. From UCSB English Broadside Ballad Archive. https://ebba.english.ucsb.edu/ballad/21317.

The Gelding of the Devil. / OR, / The Prittiest Jest That e're Was Known, / How the Bakers Wife Her Skill Hath Shown, / Then Listen a While, and I the News Will Tell, / Betwixt the Baker and the Devil of Hell. 1670. Document. British Library – Roxburghe 2.192–93.

The Hartford-Shire Mens / Fears of the Maidens Furies. / It Being an Answer to the Nine Maidens Attempt in Gelding the / Young-Man. Document. London: James Bissel. 1685–1688. From UCSB English Broadside Ballad Archive. http://ebba.english.ucsb.edu/ballad/21290.

The Holinshed Project. 2017. The Holinshed Texts (1587). http://english.nsms.ox.ac.uk/holinshed/texts.php?text1=1587_5006. Accessed 13 January 2017.

The Ladies Advocate: Or, Wit and Beauty a Match for Treachery and Inconstancy. 1749. Printed for C. Long.

The Miserable Maulster: Or, The Kentish Man's Cruelty. Who, in a Passion Dismember'd Himself, to the Unspeakable Grief of his Loving Wife, and Kind

Friends. n.d. Document. London: J. Raven. From UCSB English Broadside Ballad Archive. https://ebba.english.ucsb.edu/ballad/33671.

The Nine Maidens Fury / To The / Hartford-Shire Man. / Who Would Have Gelded Him, but That His Sweetheart Susan Coming in, Begged His Pardon. 1685–1688. Document. London: James Bissel. From UCSB English Broadside Ballad Archive. https://ebba.english.ucsb.edu/ballad/21289.

The Priest Gelded: Or Popery at the Last Gasp. Shewing … the Absolute Necessity of Passing a Law for the Castration of Popish Ecclesiastics. 1747. A. M'Culloh.

The Surgical Section. 1692. Manuscript. Wellcome Library MSMSL3/2.

Throckmorton, Anne Ruth. 2022. The Persistence of Saints in Early Modern England. PhD diss., University of Virginia. ProQuest. https://www.proquest.com/docview/875288891/abstract/B4F2C654F836458DPQ/1. Accessed 17 November 2022.

Tilmouth, Christopher. 2007. *Passion's Triumph over Reason: A History of the Moral Imagination from Spenser to Rochester.* Oxford University Press.

Tischler, Victoria. 2020. Health Humanities and the Creative Disciplines. In *The Routledge Companion to Health Humanities*, ed. Paul Crawford, Brian Brown, and Andrea Charise. Routledge.

Toulalan, Sarah. 2007. *Imagining Sex: Pornography and Bodies in Seventeenth-Century England.* Oxford University Press.

Toy, John. 1642. *A Sermon Preached in the Cathedrall Church of Worcester the Second of Febr. Last Being Candlemas Day, at the Funerall of Mris Alice Tomkins Wife unto Mr Thomas Tomkins One of the Gentleman of His Majesties Chapell Royall.* S.N.

Traub, Valerie. 2013. The New Unhistoricism in Queer Studies. *PMLA* 128 (1): 21–39. https://www.jstor.org/stable/23489260.

Tribble, Evelyn. 2011. *Cognition in the Globe: Attention and Memory in Shakespeare's Theatre.* Palgrave Macmillan US.

Tromly, Fred B. 1998. *Playing with Desire: Christopher Marlowe and the Art of Tantalization.* University of Toronto Press.

Tryon, Thomas. 1689. *A Treatise of Dreams & Visions Wherein the Causes, Natures, and Uses, of Nocturnal Representations, and the Communications Both of Good and Evil Angles, as Also Departed Souls, to Mankind Are Theosophically Unfolded.* S.N.

Turner, Daniel. 1741. *The Art of Surgery: In Which Is Laid Down Such a General Idea of the Same, as Is … Confirmed by Practice, … In Two Volumes. The Sixth Edition, Corrected.* Printed for C. Rivington, and J. Clarke.

Turner, Timothy A. 2013. Torture and Summary Justice in 'The Spanish Tragedy'. *Studies in English Literature, 1500-1900* 53 (2): 277–292. https://www.jstor.org/stable/24509999.

Turner, Brianna J., Sean M. Butler, Alexander L. Chapman, and Brianne K. Layden. 2013. How Often, or How Many Ways: Clarifying the Relationship Between

Non-Suicidal Self-Injury and Suicidality. *Archives of Suicide Research* 17 (4): 397–415. https://doi.org/10.1080/13811118.2013.802660.

Tyrell, James. 1696. *The Generall History of England, as Well Ecclesiastical as Civil.*

Vandereycken, Walter, and Ron van Deth. 1994. *From Fasting Saints to Anorexic Girls: A History of Self-Starvation.* The Athlone Press.

Vizgaitis, Alexandra L. 2015. Pierced Identities: An Investigation of the Relationships Between Body Modification Practices, Borderline Personality Pathology, and Identity Diffusion. PhD diss. ProQuest Dissertations Publishing. https://www.proquest.com/dissertations-theses/pierced-identities-investigation-relationships/docview/1705859859/se-2 (accessed November 14, 2025).

Wanley, Nathaniel. 1673. *The Wonder of the Little World, Or, a General History of Man.* Printed for T. Basset, R. Cheswel, J. Wright, and T. Sawbridge.

Watts, William. 1637. *Mortification Apostolicall.* John Legat for John Cowper.

Webster, Jeremy W. 2008. Queering the Seventeenth Century: Historicism, Queer Theory, and Early Modern Literature. *Literature Compass* 5 (2): 376–393. https://doi.org/10.1111/j.1741-4113.2008.00527.x.

Wells, Marion A. 2002. 'To Find a Face Where All Distress Is Stell'd': 'Enargeia', 'Ekphrasis', and Mourning in 'The Rape of Lucrece' and the 'Aeneid'. *Comparative Literature* 54 (2): 97–126. https://www.jstor.org/stable/4122478.

Wesley, John. 2017. Original Gesture: Hand Eloquence on the Early Modern Stage. *Shakespeare Bulletin* 35 (1): 65–96. https://doi.org/10.1353/shb.2017.0003.

Westerhof, Danielle. 2007. Deconstructing Identities on the Scaffold: The Execution of Hugh Despenser the Younger, 1326. *Journal of Medieval History* 33 (1): 87–106. https://www.tandfonline.com/doi/full/10.1016/j.jmedhist.2007.01.006.

Westminster, Matthew. 1853. *The Flowers of History, Especially as Relate to the Affairs of Britain.* Translated by C. D. Yonge. Henry G. Bohn.

Wheeler, Bonnie. 1997. Origenary Fantasies: Abelard's Castration and Confession. In *Becoming Male in the Middle Ages*, ed. Jeffrey Jerome Cohen and Bonnie Wheeler. Garland Publishing.

Willis, Thomas. 1679. *Pharmaceutice Rationalis: Or, An Exercitation of the Operations of Medicines in Humane Bodies.* T. Dring, C. Harper, and J. Leigh.

Willis, Deborah. 2002. 'The Gnawing Vulture': Revenge, Trauma Theory, and 'Titus Andronicus'. *Shakespeare Quarterly* 53 (1): 21–52. https://www.jstor.org/stable/3844038.

Wilson, John. 1608. *The English Martyrologie Conteyning a Summary of the Liues of the Glorious and Renowned Saintes of the Three Kingdomes, England, Scotland, and Ireland. Collected and Distributed into Moneths, After the Forme of a Calendar, According to Euery Saintes Festiuity. Wherunto Is Annexed in the End a Catalogue of Those, Who Haue Suffered Death in England for Defence of the Catholicke Cause, Since King Henry the 8. His Breach with the Sea Apostolicke, Vnto This Day. By a Catholicke Priest.*

Wolfall, Thomas. 1641. *The Doctrine and Practice of Mortification.* T.C. for John Sweeting.

Wright, Thomas. 1604. *The Passions of the Minde in Generall. Corrected, Enlarged, and with Sundry New Discourses Augmented.* Valentine Simmes and Adam Islip for Walter Burre and Thomas Thorpe.

Wright, Jonathan. 2017. The Undoing of a Man: Cuckoldry in Heywood's Edward IV. *Ben Johnson Journal* 24 (2): 246–267. https://www.euppublishing.com/doi/10.3366/bjj.2017.0196.

Xiao, Xinyao. 2018. Oxymoronic Ethos: The Rhetoric of Honor and Its Performance in Shakespeare's Julius Caesar. *Philological Quarterly* 97 (3): 263–285.

Yamamoto-Wilson, John. 2013. *Pain, Pleasure and Perversity: Discourses of Suffering in Seventeenth-Century England.* Ashgate Publishing.

Yonge, James. 1679. *Currus Triumphalis, è Terebinthô. Or An Account of the Many Admirable Vertues of Oleum Terebinthinæ.* Printed for J. Martyn.

Zahavi, Dan. 2014. *Shame, Self and Other.* Oxford University Press.

Index[1]

[1] Note: Page numbers followed by 'n' refer to notes.

A. Skuse, *Hurt Feelings*, Palgrave Studies in Literature, Science and Medicine, https://doi.org/10.1007/978-3-032-04565-2